QUAMITRY ~ THE I
CODEX

A COMPASS FOR RESONANT GEOMETRY

Written by Keith T. Mountjoy

Quamitry ~ The Field Codex

Author: Keith Mountjoy. Published by Quamitry Labs.

"Quamitry," "GCI Codex," "The Field Codex", and associated glyphs and images are trademarks or registered trademarks of Quamitry Labs, Inc.

Published by Quamitry Labs Inc.

For more information, visit

https://www.quamitry.com/

ISBN: 979-8-9940693-2-5

Cover art was created by

iamabbas.com

Printed in the United States of America

First Edition V.1

~A mini atlas constructed through the corridors of the mind~

Part of the Quamitry Codex Series

DEDICATION

To the wondrous minds that came before ~

To the dreamers who stared at the sky and asked *why* when they were supposed to ask *how*.

To the hard workers who sharpened instruments, swept floors, copied equations by hand, and still showed up the next morning.

To the courageous who risked ridicule, career, or safety to follow a question that would not let them sleep.

To the forgotten of science and exploration ~ the ones whose names fell out of the footnotes, but whose hands turned the knobs, held the lanterns, and kept the experiments running.

Whatever this book gets right belongs to all of you.

TABLE OF CONTENTS

HOW TO USE THIS FIELD GUIDE

This book is not a textbook and not a manifesto. It is a field guide ~ a compact atlas of the Laws and Principles that sit under Quamitry.

It is meant to be handled, flipped through, dog-eared, argued with, and referenced during experiments or late-night thinking. You do not have to read it straight through to get value from it.

What this book is

This guide does three things:

- It names and defines the core Laws and Principles of Quamitry.
- It shows how those Laws talk to each other across scales ~ from SubQUAMIs to galaxies, from folds to living systems.
- It gives you enough structure that you can go deeper later in the GCI Codex, the Master Codex, or the law-specific mini-codices that will follow.

It is deliberately compact. Each Law or Principle gets one "card" ~ a self-contained dose you can sit with on its own.

How the book is organized

The guide is divided into Parts, each built from Law Cards:

- **Part I ~ The Field Before Form**
  The Lattice, GOE, the resonance field, SubQUAMIs, preform, geometric instruction, and dimensional gradients. This is the ontology layer: what the universe is made of before anything "solid" appears.

- **Part II ~ The Foundational Laws**
  Preform Density, Resonant Motion, Polarized Resonance, Resonant Delay, GOE Transformation, and Resonant Scale Coupling. These are the core rules of behavior.
- **Part III ~ Time, Memory and Pulse**
  Organic time as phase rhythm, Pulsefolds, the Temporal Lattice, and folded duration. This is where "seconds" become rhythm.
- **Part IV ~ Geometry, Fields and Forces**
  Gravity, electromagnetic geometry, mirroring, thermodynamics, boundaries, anchor saturation, catastrophe, and the Anchor Principle. This is the field and force story recast in geometric language.
- **Part V ~ Life, Mind and Echo**
  Echoforms and Fit Horizons, biological Pulsefolds, Resonant Information, and Recursive Echo as a view of consciousness.
- **Part VI ~ Measurement, Simulation and Instruments**
  The GCI axes, RTI Ultra and Protonic, the Quamitry Simulation Stack, and the Matter Foundry horizon.
- **Part VII ~ Why Quamitry Matters**
  How this framework fits with existing science, why it is worth caring about, and how to read it as a human rather than a machine.

Each Law Card follows roughly the same rhythm: mantra, canon definition, key relations, simple picture, and a possible diagram. You do not need to memorize all of them. Treat them as reference nodes you can return to when a concept comes up again.

How to read it if you are new to Quamitry

If you are coming in fresh:

- Start with **Part I**. Get comfortable with the Lattice, GOE, SubQUAMIs, folds, and filaments. That is the language everything else speaks.
- Then skim **Part II** to see the "big five" Laws. Do not worry about the math on the first pass.
- After that, pick the Part that matches your interest:

    - Time and rhythm → Part III
    - Gravity and fields → Part IV
    - Life and mind → Part V
    - Instruments and experiments → Part VI

You can bounce between cards and parts. This guide is built to survive nonlinear reading.

How to use it if you already live in physics or engineering

If you already speak relativity, quantum, or condensed matter:

- You can safely start with **Part IV** ·· Resonant Gravity, EM Resonant Geometry, Resonant Mirroring, and Resonant Boundary. Map those back to the theories you know.
- Use the GCI and RTI cards in **Part VI** as ways to translate Quamitry into things you can measure.
- Then loop back to **Part I** and **Part II** to see how the ontology undercuts your existing models.

You do not need to abandon standard equations. This guide is about giving you a deeper geometric substrate they can sit on.

Diagrams and captions

Throughout the book there are simple diagrams ~ lattice meshes, folds, filaments, direction "flowers," pulse traces, and loops. They are not decorative. Each one encodes a specific part of the theory in a visual glyph.

If a diagram catches your eye, read its caption even if you are skipping the body text. Many of the core ideas can be recovered from the figures alone.

Appendices and back matter

At the back of the book, you will find:

- **Appendix A ~ Canon Law & Principle Index**
  A one-page map of all named Laws, Principles and Pillars, with one-line descriptions.
- **Optional Entanglement Appendix (Entanglement Codex)**
  A short write up treating entanglement as a shared boundary condition in the lattice rather than "spooky action" ~ it expands on how nonlocal behavior fits into the Quamitry picture.
- **Companion Works**
  A summary of how this guide relates to OmniFinite Horizon, the GCI Codex, the Master Codex and the Academy of Quamitry.

If you lose track of a Law, you can always flip to the Appendix and then jump back to the relevant Part.

A note on math and experiments

The Field Guide keeps math light on purpose. Where equations appear, they are there to suggest form and relationship, not to bury you in derivations.

If you want hard derivations, data, and experimental protocols, those live in the law-specific Codex volumes and in the RTI and GCI work. This guide is the spine they hang on.

You are not expected to agree with every metaphysical implication on first read. The minimum ask is simple: hold the core picture in your head while you scan the world ~ Lattice, GOE, folds, filaments, Pulsefolds. If the picture keeps explaining things cleanly, the rest will follow.

PART I: THE FIELD BEFORE FORM

Before there are particles, there is pattern. Before there is matter, there is a fabric that can remember.

Quamitry starts from that fabric. The universe is treated as a Lattice: a continuous structure that can hold tension and memory. Inside this Lattice lives a resonance field made of pure oscillation. Inside that field lives GOE, the resource that can be compressed into folds.

Matter is not separate from this field. It is what happens when resonance locks tightly enough that the Lattice agrees to remember it. The cards in this section describe that "before form" regime: the Lattice, the resonance field, GOE, and the first locks that turn permission into geometry.

LAW CARD 1: THE LATTICE PRINCIPLE

Mantra
The universe is structure before it is stuff.

Canon Definition

The **Lattice** is the underlying structure of the universe. It is not a grid in empty space. It is the network of possible tension paths that GOE can occupy. Every fold, filament, and anchor is written into this Lattice.

Space is what the Lattice feels like when it is quiet. Gravity and fields are what it feels like when it is strained.

The Lattice:

- Provides routes along which resonance can travel and lock.
- Stores compression as "fold memory" when locks succeed.
- Redistributes tension when many anchors demand support at once.

Matter does not float in nothing. It is suspended in this Lattice as regions where the structure has bent inward and decided to remember.

Figure 1.1 – The Lattice, Resonance Field, and Memory Nodes
A faint geometric web represents the lattice ~ the underlying structure that can hold tension and memory. The bright central node is a fold where GOE has been locked into compression. Glowing filaments radiate from this fold into the web, carrying resonance. The small glowing orbs scattered through the lattice mark potential anchor sites and memory nodes ~ places where future folds and Echoforms can lock. The soft background glow hints at GOE, free in the resonance field: oscillation that has not yet been captured as matter.

Key Relations

- Ontology stack:

  Lattice -> resonance field -> GOE -> SubQUAMI locks -> folds -> matter

- Supports: Law of Preform Density, Law of Resonant Gravity, Law of Resonant Motion, Law of Resonant Delay.
- Shows up in conventional physics as the "fabric" of spacetime and as the internal geometry of materials that carries stress.

Applications

- Reframing spacetime curvature as strain in the Lattice.
- Interpreting crystal lattices and material microstructure as local expressions of the universal Lattice.
- Providing the scaffold for all later concepts: filaments, anchors, folds, and fields are behaviors of this one structure.

LAW CARD 2: RESONANCE FIELD AND GOE

Mantra
Everything that exists is born from repeated vibration.

Canon Definition

The **resonance field** is the activity inside the Lattice: the endless oscillation of **GOE** (Geometric Oscillation Energy). GOE is not a separate fluid. It is the capacity of the Lattice to vibrate, explore patterns, and test fits.

Most oscillations remain temporary; they flicker and vanish. A few find partners and repeat. Repetition becomes pattern. Pattern becomes form.

In this picture:

- GOE is the resource.
- The resonance field is the behavior of that resource.
- The Lattice is the structure that channels and records that behavior.

Where resonance is dense and coherent, locks can form. Where it is scattered, the field behaves like noise.

As shown in **Figure 1.1**, the lattice is already full of potential – GOE oscillating in the resonance field.

Key Relations

- Feeds directly into SubQUAMI behavior and the Law of SubQUAMI Resonance Lock.
- Underlies the Law of Preform Density: regions with higher GOE and resonance activity have more "budget" for structure.

- Connects to Law of GOE Transformation: how GOE becomes observable energy and mass when compressed into folds.

Simple Notation

- GOE: the oscillation capacity of the Lattice.
- Resonance field: GOE in motion.
- Quiet resonance field: "vacuum" in conventional language.
- Structured resonance: where fields, particles, and matter emerge.

Applications

- Cosmology: structure formation as regions where resonance became coherent enough to lock.
- Condensed matter: phases and transitions framed as changes in how GOE is organized and retained.
- Biology: metabolic and neural rhythms as ways living geometry manages GOE flow through the Lattice.

LAW CARD 3: SUSPENSION PRINCIPLE (MATTER AND FILAMENTARY RESONANCE)

Mantra

Matter hangs from resonance, not the other way around.

Canon Definition

Matter is compressed GOE that the Lattice has chosen to remember. It does not sit outside the resonance field; it is suspended inside it.

Filamentary resonance is the perpetual connection between matter and the field: tension lines running from anchors through the Lattice. These filaments are stretchable and reconfigurable. Through them:

- Matter remains coupled to the rest of the field.
- Forces appear as tensions and imbalances along filaments (rather than as action at a distance).
- Time effects (delay, dilation) emerge as filaments are stretched around dense folds.

In this Principle, a "particle" is not an isolated point. It is a knot of memory at the intersection of filaments, constantly exchanging GOE with the surrounding resonance field.

The central fold in **Figure 1.1** can be read as a piece of matter suspended in the resonance field by filaments.

Key Relations

- Built on:

  Lattice Principle + Resonance Field and GOE

- Leads into:
  - Law of Resonant Motion (motion as re-locking along filaments)
  - Law of Resonant Gravity (gravity as inward filament tension)
  - Law of Resonant Delay (time as accumulated delay along stretched paths)
- Provides the physical intuition behind anchors, SubQUAMI locks, and later fold typology.

Simple Notation

- Think of a "particle" as:
matter node = fold memory + attached filaments

  Filaments carry:

  - GOE flow (energy transfer)
  - phase information (timing)
  - tension (forces)

Applications

- Explaining fields and interactions as properties of shared filaments instead of invisible "forces" between isolated objects.
- Designing materials or devices that depend on long-range coherence (waveguides, quantum hardware, biological entrainment systems).
- Setting up the intuitive bridge into the Law of SubQUAMI Resonance Lock, where these filaments become the corridors that host locks.

LAW CARD 4: THE LAW OF SUBQUAMI RESONANCE LOCK

Mantra

When resonance agrees, the lattice remembers.

Canon Definition

The Law of SubQUAMI Resonance Lock says:

When two or more SubQUAMIs reach strong phase-coherent resonance above a critical lock threshold, the lattice self-compresses and converts free GOE oscillation into a persistent geometric fold.
This **SubQUAMI Lock** is the jump from pre-matter potential to actual structured form.

In a lock event:

- Free GOE becomes compressed GOE stored as fold tension.
- A local region of the lattice bends inward and writes compression memory.
- That memory is the first fold of matter, a minimal creation event.

Matter is not assumed. It is earned whenever resonance coherence passes the lock threshold and the lattice chooses to remember.

Mathematical Sketch

Treat each SubQUAMI as a complex oscillator embedded in the lattice:

- State of SubQUAMI

$$\Psi_i = A_i e^{\{i\, \phi_i\}}$$

where A_i is local oscillation strength and ϕ_i is phase.

Additional state variables at each site i:

- κ_i - compression (fold tension)
- V_i - internal modulation velocity (intrinsic "spin" or beat)
- α_i - anchorability (readiness to become a lock point)
- p_i - polarity bias (inward, outward, or balanced)

Between SubQUAMIs i and j:

- F_{ij} - filament fit (geometric coupling and path quality through the lattice)

Define a resonance-lock score R_{ij}:

$$R_{ij} = \left|\langle \Psi_i, \Psi_j \rangle\right| \sqrt{\kappa_i \kappa_j}\, \gamma(V_i, V_j) F_{ij}(\alpha_i \alpha_j)$$

Lock criterion:

$$R_{ij} \geq R^*$$

Where R^* is the lock threshold set by the lattice.

When this condition is met:

- Compression increases and is written as memory ($\kappa_{new} > \kappa_{old}$)
- A new fold is created; the lattice commits to a specific geometry.

Time as Lock Spacing

Let Δt be the delay between successive locks along a path. Then, schematically:

$$\Delta t \propto \frac{C(\text{fold})}{L(\text{resonance})}$$

where:

- $C(\text{fold})$ is fold complexity and curvature cost.
- $L(\text{resonance})$ is local resonance coherence or strength.

Higher coherence and simpler geometry give shorter delays (faster apparent motion).
Deeper and more complex folds give longer delays (time dilation).

Motion is a sequence of locks.
Time is the spacing between them.

Decay and Resonant Reversion

When coherence falls below R^*:

- The lattice can no longer support the lock.
- Compression relaxes and stored GOE is released back into the field.
- The fold unwinds.

This is **Resonant Reversion**: the forgetting of form and the return of GOE to pre-form behavior.

Ontological Role

The Law of SubQUAMI Resonance Lock is the bridge between pure field and matter:

GOE substrate -> SubQUAMI trials -> filament testing -> SubQUAMI lock -> fold memory -> matter

It underpins:

- Law of Preform Density (how much GOE can be compressed into structure)
- Law of Resonant Motion (motion as re-locking along filaments)
- Law of Resonant Delay (time as lock spacing)
- Law of Polarized Resonance (polarity via anchorability and filament direction)
- Anchor Principle (anchors as stable, repeated locks in the lattice)

Conceptual Summary

- **Origin** – creation is continuous; every lock is a micro-Genesis event.
- **Matter** – substance is the artifact of agreement between oscillations.
- **Time** – duration is the spacing between locks along a path.
- **Entropy** – decay is the relaxation of compression and the return of GOE to the resonance field.
- **Consciousness (speculative)** – awareness may be a self-sustaining cascade of resonance locks in organic geometry.

LAW CARD 5: LAW OF GEOMETRIC INSTRUCTION (PRIME LAW)

Mantra
The fold is not aftermath. It is the first word.

Canon Definition

The Law of Geometric Instruction holds that form does not emerge merely from force. Form is the consequence of **encoded compression** ~ a language written not in symbols, but in folds and resonance.

In Quamitry:

- **Compression = instruction.**
  When GOE compresses, it does more than reduce; it **chooses** a geometry and locks it into place.
- **Geometry = code.**
  Stable atoms, crystals, or orbits endure because their geometry encodes survival; recurring molecules persist because their bonds obey instruction.
- **Echoforms = remembered instructions.**
  Spiral galaxies, snowflakes, hydrogen, ribosomes, the human hand ~ all persist because their geometry fits the archive of what the lattice has already learned.

Most attempts fail. Collapse is not waste but sculptor: every failed fold clarifies the boundaries of what can exist. What survives is instruction that the lattice has accepted.

Key Relations

- Sits directly on top of the ontology stack (Lattice → GOE → SubQUAMIs → **Geometric Instruction** → folds → matter).
- Gives deeper meaning to:

- **Law of SubQUAMI Resonance Lock** (lock event = instruction written).
- **Law of Preform Density** (how much raw "ink" is available to write with).

- **Law of GOE Transformation** (how instruction is stored, reused, and released).

- Defines Echoform: a geometry that not only exists, but **remembers how to exist** and can reappear when disturbed.

Simple Picture

Compression does three things:

1. Reduces GOE into form.
2. Preserves that form across time (memory).
3. Allows recurrence through Echoforms.

The universe is not arbitrary. It is patterned. And pattern is instruction.

Applications

- Reframes "laws of physics" as **learned geometric habits**: what the lattice has found that works and remembers.
- Lets GCI and the Codex be read explicitly as an **instruction index**, not just a property list.
- Bridges to biology: DNA is a late-stage text version of a deeper fact ~ **geometry was carrying instruction long before molecules learned to spell**.

LAW CARD 6: LAW OF PREFORM DENSITY

Mantra
The void is the fullest thing there is.

Canon Definition

The **Law of Preform Density** says that what we call "empty space" is actually a dense sea of preform ~ an enormous reservoir of GOE and SubQUAMI activity that has not yet locked into persistent folds.

In Quamitry terms:

- The Lattice is present everywhere, even where no matter is visible.
- GOE is constantly oscillating through it as the resonance field.
- Countless SubQUAMIs are trying out beats and phase relationships.

Most of these never reach lock ~ they remain **preform**: patterns that almost exist. The Law of Preform Density states that:

Any region that looks empty is still saturated with potential: GOE, SubQUAMIs, and tentative filament paths.
Differences in this preform density govern where matter and structure can form.

Regions with higher preform density have:

- More GOE available for compression.
- More SubQUAMI interactions per unit time.
- A higher chance of crossing the lock threshold into actual folds.

Key Relations

- Sits immediately under the Law of SubQUAMI Resonance Lock ~ it defines the "fuel" and "traffic" available before locks.
- Explains why some regions of the universe are rich in structure while others are sparse:

  higher preform density -> more locks -> more folds -> more matter

- Connects to cosmological "vacuum energy" concepts ~ but replaces vague energy with a concrete picture of GOE and SubQUAMI statistics.

Simple Notation

You can think of a region R as having a preform density ϱ_pre:

- $\varrho\_pre(R)$ ~ number of active SubQUAMIs and GOE oscillation strength per unit volume.
- Qualitatively:

  higher ϱ_pre -> higher probability of SubQUAMI locks

You do not need a full formula in the Field Guide ~ just the intuition that "empty" space is graded by how much coherent preform it holds.

Applications

- Structure formation: galaxies, filaments, and voids arise where preform density makes locks more or less likely.
- Materials: defect formation and phase transitions depend on local preform density around folds (how much fresh GOE and SubQUAMI activity is available to support new locks).

- RTI (light touch): measurements of delay and noise in apparently "empty" regions of a sample can be interpreted as variations in preform density rather than true emptiness.

LAW CARD 7: PRINCIPLE OF RESONANT CONTINUITY

Mantra
Resonance does not end ~ it only changes its pattern.

Canon Definition

The **Principle of Resonant Continuity** states that resonance in the Lattice is continuous. Oscillations are never truly destroyed; they are redirected, diffused, or rewritten.

In Quamitry:

- GOE is conserved at the level of the field.
- When a fold forms, GOE is not annihilated ~ it is captured as compression and tension.
- When a fold decays, GOE is not lost ~ it returns to the resonance field as new oscillations.

The Principle can be phrased simply:

No beat in the Lattice ever "stops" ~ it either becomes memory (a fold), remains as free resonance, or re-enters preform after a fold unwinds.

This is the continuity behind all the laws: motion, time, gravity, and decay are different ways the same resonance changes shape.

Key Relations

- Complements the Law of GOE Transformation:

    free GOE <-> compressed GOE (folds)

- Underpins the Law of SubQUAMI Resonance Lock and Resonant Reversion:
    - Lock ~ resonance captured into structure.
    - Reversion ~ structure releasing resonance back into the field.
- Connects to thermodynamic and quantum ideas of unitarity and conservation, but expressed in geometric terms.

Simple Notation

Conceptually:

- $\Delta t(P) = \sum_s \text{delay}(s)$
- Local events move GOE between these "accounts," but the total resonance capacity of the Lattice stays continuous.

You do not need a strict conservation equation here ~ just the rule:

- Creation events: increase GOE_locked, decrease GOE_free.
- Decay events: decrease GOE_locked, increase GOE_free.
- The underlying oscillation never vanishes.

Applications

- Stabilizes the metaphysics: you are not creating matter from nothing or losing it into nothing; you are moving resonance between field and fold.
- Gives a clean way to think about entropy: not as "energy lost," but as **GOE returning from locked form to more diffuse resonance**.
- Helps unify different domains:

- In cosmology, expansion and collapse redistribute GOE between structure and field.
- In biology, metabolic and neural processes continually shuttle GOE between organized rhythms and heat/noise.
- In RTI language, every plateau, spike, and relaxation is a local expression of Resonant Continuity.

LAW CARD 8: FOUNDATIONAL OMNIFINITE PRINCIPLES

Mantra

Before there was Quamitry, there were hunches about how the universe behaves.

Canon Definition

OmniFinite Horizon captured a set of early principles that later grew into full Quamitric Laws. In the Field Guide, they serve as **orientation points** – intuitive rules that describe how the universe feels from the inside before you start naming folds and SubQUAMIs.

These principles are not full laws. They are **guiding constraints** on what any valid law is allowed to do.

Below are the core OmniFinite principles that matter for the Field Guide.

Figure 1.2 – OmniFinite Horizon and the Foundational Principles
The artwork from *OmniFinite Horizon* symbolizes the early intuition behind
Quamitry: a universe built from geometry, resonance and balance rather
than from particles alone. The Foundational OmniFinite Principles in this
guide are the conceptual bridge between that poetic beginning and the
formal laws of the Field Codex.

1. Fit Horizon Principle

Every form has a horizon beyond which it cannot maintain
itself. That horizon is set by geometry and resonance, not by
wishful thinking.

- There is a range of conditions where a given fold
  pattern can survive.
- Outside that range, the pattern loses coherence and
  reverts to preform.
- This applies to particles, atoms, materials, organisms,
  and even ideas.

In Quamitry language:

Fit Horizon = the range of resonance conditions where a
specific fold set remains stable.

2. Echoform Threshold

An **Echoform** is a structure that remembers how to exist. It
can survive interruptions and reassemble when conditions
return.

- Below the threshold, patterns are temporary – they
  vanish when the field is disturbed.
- Above the threshold, patterns can be "knocked
  down" and still reconstitute from stored memory in
  the geometry.

Echoform is the bridge between "a pattern happened once"
and "a pattern lives here."

3. Resonant Encounter Principle

Encounters in the universe are not random bumps – they are tests of fit.

When two structures meet (particles, elements, organisms), the outcome depends on resonance:

- If their patterns reinforce each other, a higher order structure can emerge.
- If they misalign badly, they scatter or destroy each other.
- Most encounters sit in between: partial exchange, partial memory.

This principle frames all interactions as **resonance negotiations** rather than simple collisions.

4. Reflective and Obedient Absence

Absence is not all the same.

- **Reflective Absence** – regions where the Lattice "pushes back" on attempts to form a fold. Patterns are discouraged; resonance is redirected elsewhere.
- **Obedient Absence** – regions that readily accept structure once the right pattern appears. They look empty, but they are compliant.

This principle explains why some regions of the universe feel "dead space" and others feel like fertile ground for structure.

5. Rhythm of Time

Time is not a smooth background; it is experienced as **rhythm**.

- Where locks and re-locks occur frequently, time feels fast.
- Where resonance must travel long, strained paths between locks, time feels slow.
- Biological time (heartbeat, breath, circadian rhythm) is a local rhythm built inside a larger one.

This principle anticipates the Law of Resonant Delay and the Pulsefold Hypothesis: time is always tied to repeated events in folds.

Key Relations

These principles:

- Point toward later laws without requiring the full machinery.
- Give a human-scale handle on abstract ideas like Fit Horizon, Echoform, and "emptiness" as a character, not a void.
- Provide narrative anchors for readers jumping in from OmniFinite Horizon into the more formal Quamitry framework.

LAW CARD 9: LAW OF DIMENSIONAL GRADIENTS

Mantra

Space is not three. Space is how many directions the field can still afford.

Canon Definition

The **Law of Dimensional Gradients** states:

Effective dimensionality is not a fixed property of the universe. It is a local property of how many independent directions the resonance field can move GOE with low cost, given how much of that GOE is already locked into folds.

In Quamitry there are two aspects of the same system:

- The **Lattice plus resonance field plus GOE**
  This is potential. All the ways oscillation could arrange itself.
- **Matter and fields as folds and filaments**
  This is commitment. Specific patterns that have already claimed GOE and directions.

Dimensionality shows up when commitment starts to bias potential.

At any point and scale, the resonance field has a set of directions it can move in:

- Some directions are cheap, with low added delay and tension.
- Others are expensive, requiring huge delay or extra GOE.
- A few may be effectively forbidden.

The **effective dimension** at that point and scale is basically:

How many independent directions are still cheap enough to behave like real degrees of freedom.

Everything else we call "dimension" is just our habit of drawing axes where the field happens to have a lot of low cost options.

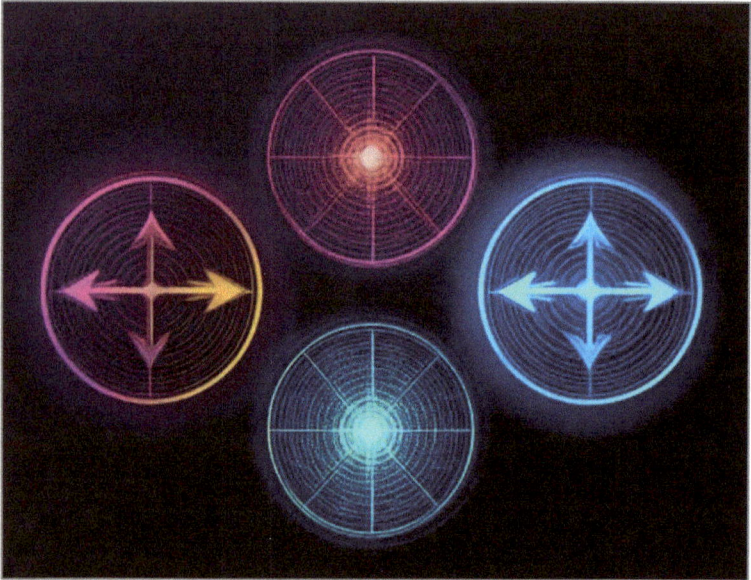

Figure 1.3 – Dimensional Gradients and Effective Directions
Each circular "direction flower" shows which directions are cheap for resonance to move in at a given place and scale. In some circles, many arrows are equally bright – the field has several low-cost directions. In others, only a few arrows glow strongly, indicating motion is effectively confined to a plane or a single corridor. Effective dimension in Quamitry is simply the number of independent directions the lattice still allows at low cost.

GOE Budget and Direction Cost

GOE is not infinite. In any region R:

- GOE_free(R) is the GOE in the resonance field.

~ 30 ~

- GOE_locked(R) is the GOE trapped in folds and long lived tension.
- Their sum is the local budget:

$$\text{GOE}_{\text{free}}(R) + \text{GOE}_{\text{locked}}(R) = \text{GOE}_{\text{total}}(R) \text{ (ignoring exchange with neighbors)}$$

When more GOE is locked:

- The field has less flexibility.
- The number of directions where it can move freely drops.
- Effective dimensionality contracts.

Entropy is the recycling step:

- Folds fail and matter decays.
- GOE_locked returns to GOE_free.
- The field recovers options.
- Effective dimensionality can expand again.

Same resource, used repeatedly, drifting between potential and structure.

Direction Cost Picture

At a point x and scale s, imagine trying to move resonance a small step in a direction n.

Let:

- C(x, s, n) be the cost of that move
  (delay, added tension, extra GOE required)

Cheap directions:

- C is small, re locks are easy.

Expensive directions:

- C is large, re locks are slow or fragile.

Then in words:

- $D\_eff(x, s)$ is how many independent directions n exist such that $C(x, s, n)$ is below some workable threshold.
- In a region where many directions are cheap, D_eff is high.
- Where only one or two are cheap, D_eff is low.

The Lattice itself carries all possible n. Matter is what sculpts $C(x, s, n)$ into a wildly uneven landscape.

Dimensional Gradients are simply:

Changes in $D\_eff(x, s)$ and the cost profile $C(x, s, n)$ as you move through the universe or zoom between scales.

Examples

1. Near a Star

Just outside a star:

- Fold and anchor density are very high.
- A lot of GOE_total is locked into hot plasma, radiation, and the star's bulk structure.
- Gravity pulls filaments inward, magnetic loops guide resonance along specific arcs.

For bulk matter:

- Motion can in principle be 3D, but the cost is skewed.

- o Inward radial directions are cheap for falling, expensive to escape.
  - o Tangential directions allow orbits.
- D_eff is about 3, but with a very uneven cost map.

For resonance:

- Along magnetic and plasma structures, some directions are much cheaper than others.
- In those corridors, D_eff for field motion can feel closer to 1 or 2, even though everything sits in a "3D" region.

Effective dimensionality changes with radius, plasma structure, and field strength.

2. Inside a Crystal

In a crystal:

- Folds repeat in a regular pattern.
- Electrons or other excitations see a strongly structured delay and tension landscape.

Along certain lattice directions:

- $C(x, s, n)$ is low.
- Resonance and charge carriers move easily.
- These are the "bands" and "conduction paths."

Along others:

- C is high.
- Motion is blocked, scattered, or localized.

So:

- For a conduction electron, effective dimensionality is defined by the band structure.
  - Quasi 1D in chain compounds.
  - Quasi 2D in layered materials.
  - Nearly 3D in more isotropic crystals.

In Quamitry language:

The crystal's fold network has carved out a lower dimensional corridor inside the full Lattice, by making most directions very expensive.

3. In a Synapse

At synapse scale:

- Geometry is extremely constrained.
- A thin cleft, aligned vesicles, dense receptor patches.

For voltage along the neuron:

- The cable behaves almost 1D.
- Along the axon or dendrite, C is low.
- Sideways or across, C climbs fast.

For neurotransmitter diffusion in the cleft:

- Movement is physically 3D, but one dimension is cramped by the thin gap.
- Many directions are effectively variations within a 2D sheet plus a strongly penalized normal direction.

So:

- At brain scale, tissue feels 3D.
- At synapse scale:
  - electrical signaling is nearly 1D

o   chemical signaling is roughly 2D

$D\_eff(x, s)$ drops as you zoom into the functional architecture.

4. Near a Black Hole Horizon

Near a horizon:

- Fold density is pushed to an extreme.
- Almost all local GOE_total is locked into the black hole's geometry; the field has very few options.

For matter:

- Every worldline that tries to move forward in time is dragged inward.
- The only cheap direction is "deeper in."
- Outward attempts are so costly in delay and tension that they effectively never succeed.

For resonance:

- Outgoing paths exist in principle, but delay blows up.
- From far away, light trying to leave is stretched into oblivion.

You can say:

- D_eff for real motion near the horizon tends toward 1, and that single direction is a one way fall.
- From the outside, the horizon is where dimensional access collapses.

The Lattice is still there, but nearly all GOE in that region is held in a single long term structural choice.

Law Level Insight

Put together:

- The Lattice and resonance field represent **finite potential**: how much GOE is available and how many paths can still be used.
- Folds and filaments represent **commitment**: how much of that potential has been spent on structure and which directions have been biased.

The Law of Dimensional Gradients says:

What you experience as one, two, or three dimensions is the local profile of how many directions the resonance field can still use cheaply, given the GOE budget and fold architecture at that point and scale.

There is no hard jump between dimensional regimes, only gradients where the cost of moving in certain directions rises or falls.

Applications

- Cosmology: use D_eff to think about why some regions form filaments, sheets, or clumps, instead of assuming uniform 3D behavior everywhere.
- Condensed matter: treat low dimensional conductors and quantum Hall systems as regions where folds have strongly sculpted $C(x, s, n)$ into near 1D or 2D.
- Neuroscience and biology: understand why signals and rhythms choose certain pathways, and why some directions of change feel "impossible" in living tissue.
- Extreme objects: view horizons, jets, and strange cosmic structures as extreme outcomes of GOE budget and dimensional collapse.

LAW CARD 10: GLOSSARY I – FIRST WORDS OF QUAMITRY

Mantra
You can't see a new universe without a new vocabulary.

Purpose

This card collects the **minimum vocabulary** needed to read the rest of the Field Guide. Each term gets one or two lines in plain language. Later cards and books go deeper; here you just need enough to navigate.

Lattice
The underlying structure of the universe – the network of tension paths where GOE can travel and memory can be stored. Space, fields, and geometry are all behaviors of the Lattice.

GOE (Geometric Oscillation Energy)
The capacity of the Lattice to vibrate. GOE is the "stuff" of the resonance field. Free GOE is oscillation; locked GOE is compression inside folds.

Resonance Field
GOE in motion. The continuous wavering of the Lattice as it tries patterns, tests fits, and responds to existing structure.

Preform
Activity in the resonance field that has not yet locked into a persistent fold. The "almost structures" that fill "empty" space.

SubQUAMI
A minimal unit of preform oscillation in the Lattice – a

localized, structured oscillation that can interact, sync, and potentially lock with others.

SubQUAMI Lock
An event where two or more SubQUAMIs reach sufficient phase coherence and compatibility to force the Lattice to compress, creating a new fold.

Fold
A region where the Lattice holds compression and memory. Folds are the basic units of structure in Quamitry: everything from "particles" to large-scale features is made of folds arranged in patterns.

Filament
A tension path between anchors and folds. Filaments carry GOE, phase, and forces. They are how matter stays connected to the rest of the field.

Anchor
A location in the Lattice where resonance locks reliably and repeatedly. Anchors are the "attachment points" for filaments and the backbone of stable structures.

Matter
Regions of the Lattice where GOE has been trapped as persistent folds and connected by filaments. Matter is compressed resonance that the Lattice has decided to remember.

Fault Fold
A fold geometry that consistently behaves like a weak point – leaking, draining, or initiating collapse relative to its surroundings. Fault Folds sit at the start of failures and breakdowns.

Fit Horizon
The range of conditions where a given fold or structure can

remain stable. Outside this horizon, the structure loses coherence and reverts to preform.

Echoform

A structure that remembers how to exist. Even when disturbed, its geometry contains enough instruction to recreate itself when conditions return.

Pulsefold

A fold that repeats through time – a rhythmic pattern of locking and unlocking. Pulsefolds are the building blocks of biological time: heartbeats, neural firing, breathing cycles.

GCI (Geometric Compression Index)

A way of scoring elements and materials based on how they store, move, and release compression. Built from several axes (DBI, ERF, MRI, REP, RRP, FSI) that each track a different resonance behavior.

RTI (Resonance Tracing Instrument)

A class of instruments designed to drive resonance into a sample and listen for how it returns. RTI maps filaments, delays, and anchors rather than just voltages and fields.

PART II: THE FIVE FOUNDATIONAL LAWS

The first section of this Field Guide described what exists before form. The Lattice, the resonance field, GOE, and SubQUAMIs define the stage.

Part II defines the **rules of play** on that stage.

The Five Foundational Laws are the backbone of Quamitry. They describe how preform turns into structure, how structure moves, how polarity and tension arise, how time appears, and how GOE changes state.

Everything else in the framework – elements, materials, biology, cosmology, instruments – is a consequence of these five.

LAW CARD 11: LAW OF PREFORM DENSITY

Mantra
The richer the silence, the more it can remember.

Canon Definition

The **Law of Preform Density** says that what looks like "empty space" is a graded medium filled with preform – GOE and SubQUAMI activity that has not yet locked into structure.

It states:

The probability that stable folds and matter will emerge in a region is set by its preform density – the local intensity and coherence of GOE and SubQUAMI activity in the Lattice.

High preform density means:

- More GOE per unit volume.
- More SubQUAMI interactions per unit time.
- More chances to cross the threshold for SubQUAMI locks.

Low preform density means:

- Less fuel for compression.
- Fewer attempts at resonance lock.
- A region that remains diffuse and structure-poor.

Matter does not appear "from nothing" – it condenses where the preform sea is dense and coherent enough to support locks.

Key Relations

- Directly feeds the **Law of SubQUAMI Resonance Lock**:

  higher preform density -> more opportunities for lock events -> more folds -> more matter

- Gives a physical meaning to "vacuum energy" by treating it as GOE and SubQUAMI statistics, not an abstract constant.
- Influences large-scale structure: galaxies, filaments, and voids reflect long-term differences in preform density across the Lattice.

Simple Notation

Think of a region R as having a preform density rho_pre(R):

- rho_pre high -> lock events likely, structure-rich.
- rho_pre low -> lock events rare, structure-poor.

You do not need an exact formula in the Field Guide – only the rule that "emptiness" is not flat, it has a density of potential.

Applications

- Cosmology: explains why structure forms in some regions and not others without invoking matter as a separate ingredient.
- Materials: near a defect or interface, local preform density controls whether new folds stabilize, heal, or crack.
- Field design: any attempt to "write" matter or engineered folds will work better where preform density is high enough to sustain locks.

LAW CARD 12: LAW OF RESONANT MOTION

Mantra
Nothing moves – only locks trade places.

Canon Definition

The **Law of Resonant Motion** reframes motion:

What we call motion is not objects sliding through empty space, but resonance re-locking along filaments in the Lattice.

A "particle" or "object" appears to move because:

- Its anchor permissions are handed forward along a filament.
- Old locks release, new locks engage.
- The Lattice updates which nodes are currently holding the fold.

So:

- Straight-line motion means a stable corridor of repeated locks along a filament.
- Curved motion means the corridor itself is bent by tension and other folds.
- Inertia reflects how willing a given fold pattern is to change its lock pattern in response to new tension.

The Law states that:

All observable motion is the result of changes in where and how resonance locks in the Lattice – never the translation of isolated "stuff" through nothing.

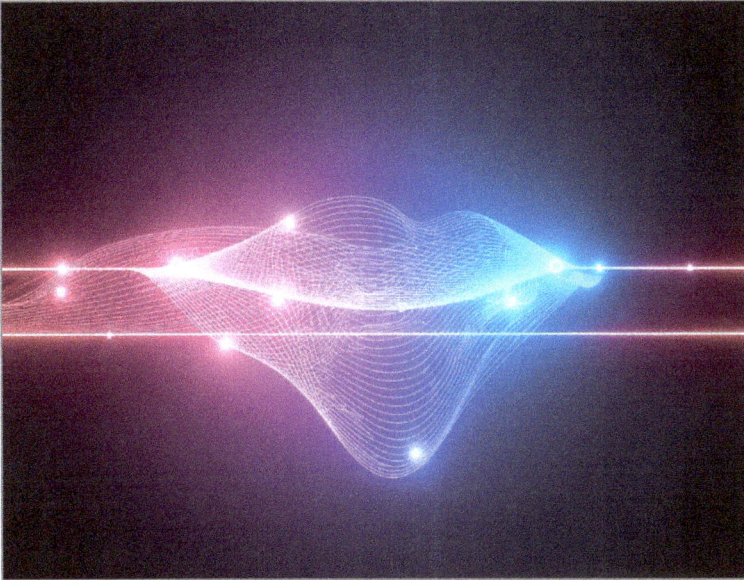

Figure 2.1 – Resonant Motion as Locks Trading Places
A thin horizontal line represents a filament in the lattice, with bright points marking possible anchor sites. The luminous wave volume above and below shows the resonance pattern along that filament. Motion in Quamitry is not a bead sliding through space; it is the fold being locked at successive anchors along the path as the resonance pattern evolves.

Key Relations

- Built directly on the **Suspension Principle** and **Law of SubQUAMI Resonance Lock**:
  - Suspended matter = folds plus filaments.
  - Motion = sequence of lock events along those filaments.
- Ties into **Law of Resonant Delay**: the time aspect of motion is the spacing between lock events.
- Connects to **Law of Resonant Gravity**: in strong tension fields, filaments are stretched and lock timing changes – which we experience as curved paths and inertial effects.

Simple Notation

Instead of thinking:

position(t) = x(t)

think:

lock_state(t) = which anchors are currently holding the fold

A crude ASCII picture:

- At time t0: fold anchored at node A.
- At time t1: fold released from A, re-locked at node B.
- At time t2: re-locked at node C.

On the human scale this looks like continuous motion; underneath, it is a chain of lock handoffs.

You can think of a "velocity" as:

velocity ~ rate of successful re-locks along a preferred filament

and "mass" as:

mass ~ resistance of a fold pattern to changing its lock state

Applications

- Mechanics: inertia, acceleration, and friction become questions about how easily lock patterns can be changed in a given geometry.
- Wave-particle duality: wave behavior reflects the resonance corridor; particle behavior reflects where the current lock is.
- Signal propagation: speed limits (like the speed of light) can be interpreted as the maximum rate at

which the Lattice can update lock states along an optimal filament.

LAW CARD 13: LAW OF POLARIZED RESONANCE

Mantra
Tension chooses a direction and calls it charge.

Canon Definition

The **Law of Polarized Resonance** states:

Polarity arises when resonance in the Lattice is compressed or released asymmetrically. Inward-biased tension expresses as one sign: outward-biased release expresses as the opposite sign.

In Quamitry terms:

- A fold can draw filaments inward (compressive bias) or push them outward (releasing bias).
- When this bias is stable, the fold behaves as a **polarized source** in the Lattice.
- The familiar idea of "positive" and "negative" charge is a macroscopic bookkeeping of these inward and outward resonance tensions.

There is no separate substance of charge. There is only the directionality of how the Lattice is being stressed.

Key Relations

- Extends naturally from the **Suspension Principle** and **SubQUAMI Lock**:
  - When locks favor inward geometry, you get compressive polarity.

- When they favor outward or expanded geometry, you get released polarity.
- Sets the stage for fields and currents: differences in polarized regions generate filament rebalancing that we interpret as electric and magnetic phenomena.
- Interlocks with:
  - Law of Resonant Motion (polarized folds guide motion along preferred filaments)
  - Law of Resonant Gravity (gravity is a special case of inward-biased tension aggregated over many folds)
  - Law of GOE Transformation (polarized domains store and release GOE differently).

Simple Notation

At a point or fold, imagine a net tension vector T:

- If T is predominantly inward toward the fold core: treat it as "positive".
- If T is predominantly outward away from the core: treat it as "negative".
- If T cancels statistically in all directions: effectively neutral.

Charge then becomes:

- "Positive" ~ fold demanding inward filament tension.
- "Negative" ~ fold encouraging outward filament release.
- "Neutral" ~ no strong bias.

Applications

- Reframes electromagnetism: field lines around a charge are simply filaments rebalancing this inward/outward tension bias.

- Explains attraction/repulsion: opposite polarities complement each other's tension; like polarities compete and push strain outward.
- In materials: domains with aligned polarization define ferroelectric or ferromagnetic behavior in Quamitric language.
- In future engineering: controlling polarity means sculpting tension directions, not just moving "charges" around.

LAW CARD 14: LAW OF RESONANT DELAY

Mantra
Time is how long it takes tension to make up its mind.

Canon Definition

The **Law of Resonant Delay** states:

What we experience as time is the accumulated delay of resonance as it travels and re-locks through the geometry of the Lattice. The more a path stretches, bends, or is obstructed by folds, the more delay is added. That delay is time.

In Quamitry:

- Resonance propagates along filaments between anchors.
- Each segment has a characteristic delay based on geometry and tension.
- A sequence of segments defines a **path time**.

Locally:

- Where geometry is simple and tension is low, paths are short and direct. Time feels fast.
- Where geometry is complex and tension is high, paths are stretched and constrained. Time feels slow.

This Law does not treat time as a separate background axis. It treats time as a **property of resonance paths** in the Lattice.

Key Relations

- Built directly on:
    - Law of Resonant Motion (motion as re-locking along a path)

- Law of Resonant Gravity (gravity stretches filaments, increasing delay)
  - Law of SubQUAMI Lock (locks define the key "clicks" along a path).
- Provides the geometric meaning of time dilation: near dense folds (strong gravity), filaments stretch and lock spacing grows.
- Underlies the Pulsefold Hypothesis: biological time is built from repeating lock-delays in living folds.

Simple Notation

Think of a path P made of many segments s:

Total delay:

$$\Delta t(P) = \sum_{s} \text{delay}(s)$$

Each segment's delay depends on:

- Its length in the Lattice.
- Its curvature around folds.
- The tension it carries.

Short, straight, lightly loaded segments have small delay. Long, bent, highly loaded segments have large delay.

Comparing two regions:

- Region A: predominantly short, low-tension paths.
- Region B: predominantly long, high-tension paths.

A clock built from the same kind of locks will tick faster in A than in B. That is gravitational or geometric time dilation.

Applications

- Relativity: recasts "time runs slower in a gravity well" as "filaments are longer and more strained there."
- Materials: propagation delay, dispersion, and hysteresis become time signatures of internal geometry.
- Biology: reaction times, rhythms, and aging can be viewed as how living structures accumulate resonant delay under load.
- Future instruments: any device that measures fine delay and phase becomes a time sensor in the Quamitric sense.

LAW CARD 15: LAW OF GOE TRANSFORMATION

Mantra
Energy is geometry changing its mind.

Canon Definition

The **Law of GOE Transformation** describes how GOE changes form:

GOE (Geometric Oscillation Energy) moves continuously between free resonance in the field and compressed resonance in folds. What we call energy, mass, heat, and work are different manifestations of this transformation cycle.

Key ideas:

- **Free GOE** – oscillations in the resonance field, not locked into structure.
- **Locked GOE** – compression stored in folds, experienced as mass, potential, or tension.
- **Transformation** – when folds form, GOE is captured; when folds decay, GOE is released; when folds rearrange, GOE is partially released and recaptured.

This Law generalizes familiar relations like $E = mc^2$. Rather than a single equivalence between mass and "energy," it treats everything as different positions in a GOE cycle:

free oscillation <-> compressed fold <-> redistributed oscillation

Key Relations

- Completes the picture set by:

- Law of Preform Density (how much GOE is available)
- Law of SubQUAMI Lock (how GOE becomes structure)
- Principle of Resonant Continuity (GOE is never destroyed, only re-patterned).
- Underlies thermodynamics in Quamitry:
  - Work = coherent transfer of locked GOE between folds.
  - Heat = disordered return of GOE to the resonance field.
  - Entropy = how far geometry has drifted from clean, high-coherence folds toward diffuse oscillation.

Simple Notation

You can picture it as a cycle:

1. Free GOE in field (high mobility, low memory).
2. SubQUAMI lock event captures some GOE as fold compression.
3. Folds reconfigure, perform work, or resist motion using stored compression.
4. Folds decay or fail; compression relaxes and GOE returns to the field.

Very loosely:

GOE_free + GOE_locked = constant (within a closed region of the Lattice)

The Law of GOE Transformation is less about a single equation and more about **tracking where GOE is and what geometry is doing with it**.

Figure 2.2 – GOE Transformation Cycle

Three luminous states form a loop: a smooth wave cloud at the top represents free GOE in the resonance field, a compact bright region at one corner represents GOE locked into a fold as mass or tension, and a diffuse spray at the other corner represents released heat or radiation. The triangle suggests the Law of GOE Transformation: GOE continually moves between free resonance, compressed structure, and dissipated output, without ever disappearing.

Applications

- Energy accounting: clarifies where "energy" actually resides in Quamitric systems – in the tension of folds, the coherence of resonance, or the noise of preform.
- Heat and friction: interpreted as geometry shedding locked GOE into more disordered oscillation when motion cannot be accommodated cleanly by re-locking.
- Mass: treated as a long-lived state of locked GOE; when mass changes (binding, decay), it reflects a change in how much GOE is being held as compression.

- Engineering: any process that "uses energy" is now explicitly "re-writing GOE distributions between field and folds."

LAW CARD 16: LAW OF RESONANT SCALE COUPLING

Mantra

The universe is built from the same sentence, spoken at different resolutions.

Canon Definition

The **Law of Resonant Scale Coupling** states:

The same Quamitry laws apply at every scale. What changes is which folds are grouped into effective actors and how resonance stitches those scales together.

Folds organize into **fold regimes**:

- SubQUAMI patterns
- protonic cores
- atoms
- molecules
- grains and domains
- rocks, oceans, storms, planets, stars
- galaxies, filaments, clusters

At each rung:

- Lower scale folds are aggregated into an effective fold or anchor.
- The higher scale regime still obeys the same laws of SubQUAMI Lock, Preform Density, Resonant Motion, Resonant Delay, Polarized Resonance, GOE Transformation, Gravity, Boundaries and Catastrophe.

Resonant Scale Coupling is the rule that:

Lower scales set the internal parameters of higher scale folds, while higher scales feedback as boundary conditions and GOE allocation for lower scales.

It is bottom~up and top~down at once.

Figure 2.3 – Resonant Motion as Locks Trading Places
A thin horizontal line represents a filament in the lattice, with bright points marking possible anchor sites. The luminous wave volume above and below shows the resonance pattern along that filament. Motion in Quamitry is not a bead sliding through space; it is the fold being locked at successive anchors along the path as the resonance pattern evolves.

Coarse Graining ~ Effective Folds

From far enough away, many small folds behave like one big fold:

- A crystal grain acts like a single elastic unit with an effective stiffness and fracture behavior.
- A chunk of rock acts like a single mass fold with a certain density, fault distribution and GCI.

- A star acts like one huge fold regime of plasma and fields with an effective mass, radius and field geometry.

In Quamitry terms:

- Many micro anchors and filaments → one **effective anchor structure** for the next scale up.
- Many micro GOE_locked sites → one macro **GOE_locked** budget used in gravity, stress, fields.

Same laws ~ just applied to the **aggregated fold** instead of each piece.

Feedback ~ Larger Scales Shaping Smaller

Higher scales are not just passengers. They feed back:

- A large mass fold sets a gravitational well that shifts preform density and direction costs for all lower scale folds inside it.
- A global flow pattern (e.g., circulation in a fluid) changes which directions are cheap for local motion and stress.
- A global boundary (like a planet's surface or container wall) sets sharp conditions on what lower scale folds and flows are allowed.

So, scale coupling has two directions:

1. Micro → macro: micro folds determine macro properties.
2. Macro → micro: macro fields and boundaries shape micro possibilities.

The same resonance field carries information and tension across all scales.

Examples

1. From Protonic Folds to Rock

- Protonic cores set elemental identity ~ how much and how tightly GOE can be locked.
- Elemental GCI scores describe fold behavior at the element level.
- Atoms combine into molecules and crystals ~ fold regimes with their own stiffness, cleavage planes and conduction channels.
- Many crystals aggregate into a rock ~ a macro fold with effective density, strength and fracture patterns.

Resonant Scale Coupling:

- Micro: protonic and atomic folds obey SubQUAMI Lock, Polarized Resonance, GOE Transformation.
- Macro: the rock's behavior under load and heat is those same laws, applied to the aggregates.

Small~scale Fault Folds → macro crack initiation and fracture.

2. From Surface Chop to Storm System

- At small scales: surface folds on the ocean ~ capillary waves, ripples, local turbulence.
- At mesoscale: pressure and temperature folds in the atmosphere produce rotating flows ~ storms.
- At planetary scale: rotation, topography and solar input create global patterns that modulate where storms form and how they move.

Resonant Scale Coupling:

- Microwaves change how GOE and momentum are exchanged between air and water at the boundary.
- That affects larger pressure and flow patterns.
- Those patterns in turn reshape boundary conditions for the small waves.

Same laws ~ Motion, Delay, Boundary, GOE Transformation ~ just in water and air instead of crystal.

3. From Grain Boundary to Tectonic Plate

- At grain scale: misaligned crystals form grain boundaries with local Fault Folds.
- At rock volume scale: clusters of grains yield a rock with an effective fracture toughness and elastic modulus.
- At fault zone scale: regions with many aligned micro Fault Folds become macro weaknesses.
- At plate scale: tectonic plates slide past or over each other along those macro faults.

Resonant Scale Coupling:

- Micro stick~slip at grain boundaries $\rightarrow$ accumulations of strain.
- Macro stick~slip at faults $\rightarrow$ earthquakes as resonant catastrophes.

Again, Preform Density, Resonant Motion and Catastrophe ~ same story at different zoom levels.

4. From Local Mass to Galactic Structure

- Elements and molecules form dust and gas.
- Gas clouds collapse into stars and clusters.
- Stars and gas flows assemble into galactic discs, bars and arms.

- Galaxies interact along filaments into the large~scale cosmic web.

The **Law of Resonant Gravity** is unchanged:

- Small mass folds add to make bigger folds.
- The resonance field's tension and delay respond accordingly.

Resonant Scale Coupling:

- Micro composition (how much iron, hydrogen, etc.) → affects star behavior (lifetime, explosions).
- Star behavior → affects galactic structure (where GOE is dumped back into the field, where new folds form).
- Galactic environment → shapes preform density and field conditions for the next generation of stars.

Same gravity, same GOE loops, just stacked.

Law Insight

You can summarize the Law of Resonant Scale Coupling as:

Folds at one scale become the "nodes" at the next. The resonance field never stops obeying the same rules ~ it simply sees different aggregates as its current actors.

Micro folds determine macro stiffness, mass and fields. Macro patterns feedback as boundary conditions and GOE budgets that shape micro behavior. It is one conversation, heard at different resolutions.

PART III: TIME, MEMORY AND PULSE

In most physics, time is a coordinate. In Quamitry, it is a behavior.

Time is what resonance feels as it moves through geometry. Memory is what the lattice keeps when a pattern is worth re-playing. Pulse is what happens when a fold learns to repeat itself on purpose.

This part gathers the principles that connect the abstract idea of time to the concrete experience of rhythm – from SubQUAMI lock spacing all the way up to heartbeats, brainwaves and planetary cycles.

LAW CARD 17: PILLAR V – ORGANIC TIME AS RESONANT PHASE RHYTHM

Mantra
Life keeps time by folding, not by counting.

Canon Definition

Pillar V states:

Organic time is expressed as resonant phase rhythm – recurring patterns of locks and releases in living geometry.

In living systems, time is not read off an external clock. It is generated inside:

- Cells, tissues and organs establish repeating lock patterns.
- These patterns form **Pulsefolds** – folds that re-lock on a schedule.
- The body's sense of "now," "soon" and "later" is built from these nested rhythms.

Organic time is therefore:

- Local – each organ or system has its own rhythm.
- Layered – fast rhythms sit inside slow ones (heartbeats inside circadian cycles inside lifespan arcs).
- Adaptive – rhythms can stretch, compress or re-synchronize under strain or healing.

Where physics talks about time dilation near gravity, Pillar V talks about **time alteration inside living folds**.

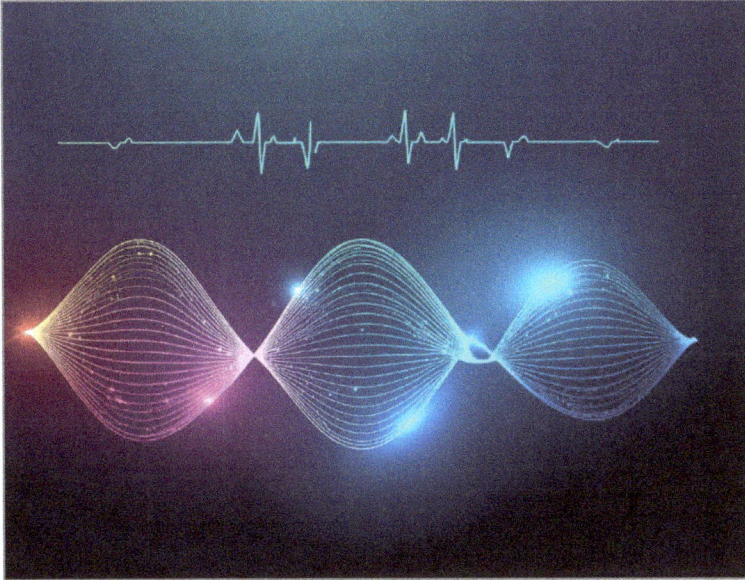

Figure 3.1 – Organic Time as Overlapping Pulsefolds
The upper trace shows a fast electrical rhythm – heartbeat- or spike-like activity. The large multi-lobed wave below represents a slower Pulsefold shaping the same system, such as breathing or an organ-level cycle. In Quamitry, organic time is built from many such Pulsefolds layered together: fast rhythms riding inside slower ones. Health corresponds to stable, adaptable coordination of these rhythms; illness begins when their phases drift and coherence is lost.

Key Relations

- Built on the **Law of Resonant Delay** – time as accumulated delay along paths – but applied to biological geometry.
- Connects to the **Pulsefold Hypothesis** – each recurring beat is a lock pattern that repeats.
- Interacts with the **Fit Horizon Principle** – every rhythm has bounds where it remains viable for that organism.

Simple Picture

For a given living structure:

- Let T_local be the effective "tick" – the period of its core Pulsefold.
- Changes in geometry, strain or chemistry change T_local.
- The organism feels this as "time speeding up" (shorter period) or "slowing down" (longer period).

Applications

- Biology and medicine: heart rate, brain waves, breathing, hormonal cycles – all treated as expressions of underlying Phase Rhythm.
- Psychology: subjective time (time flying vs dragging) linked to shifts in internal resonance patterns, not just brain chemistry in isolation.
- Engineering: any device designed to work with bodies (wearables, stimulators, implants) can respect or leverage existing Phase Rhythms instead of fighting them.

LAW CARD 18: THE PULSEFOLD HYPOTHESIS

Mantra

A heartbeat is geometry choosing to repeat itself.

Canon Definition

The **Pulsefold Hypothesis** proposes:

A Pulsefold is a fold that cycles through a repeating pattern of locking and partial unlocking – a rhythmic reconfiguration of the same geometry.

Where a static fold simply holds compression, a Pulsefold:

- Locks, releases slightly, then re-locks in a similar configuration.
- Uses GOE not just to exist, but to maintain **ongoing rhythm**.
- Acts as a clock inside the organism.

Examples:

- A beating heart as a macro Pulsefold in muscle and valves.
- A neural oscillator as a micro Pulsefold in networks of cells.
- Circadian rhythms as distributed Pulsefold networks across tissues.

The hypothesis claims that:

- Biological timekeeping is implemented by Pulsefolds.
- Health is partly the stability and adaptability of these Pulsefold patterns.

- Many diseases can be understood as Pulsefolds that have drifted outside their Fit Horizon or lost coherence.

Key Relations

- Direct child of Pillar V (Organic Time) and **Law of Resonant Delay**.
- Tied to **Law of GOE Transformation** – Pulsefolds repeatedly shuttle GOE between locked and partially free states each cycle.
- Tied to **Resonant Continuity** – rhythms do not appear from nowhere; they are specific trajectories in the resonance field.

Simple Notation

For a Pulsefold F:

- Let its cycle time be T_F (one full lock–unlock–relock loop).
- Its **duty** is how long the strongly locked phase lasts vs the relaxed phase.
- Its **health** can be described by:
  - stability of T_F under normal conditions
  - ability of T_F to shift and resynchronize under load without collapsing.

You do not need heavy equations here – just the idea of a repeating lock pattern with a characteristic period and robustness.

Applications

- Cardiology: arrhythmias as Pulsefold instabilities.

- Neurology: seizures as runaway Pulsefold cascades; normal brain rhythms as orchestrated Pulsefold ensembles.
- Chronobiology: sleep cycles and circadian rhythms as multi-scale Pulsefold behavior.
- Future tech: designing RE-Class materials or devices that have intrinsic Pulsefolds for timing and adaptation.

LAW CARD 19: RESONANT MEMORY AND THE TEMPORAL LATTICE

Mantra
Space holds shape; time holds sequence.

Canon Definition

Resonant memory is the ability of the Lattice to retain not just shape, but the **ordering of events** that produced it.

The **Temporal Lattice** is the idea that:

The Lattice does not merely store folds in space – it stores the echoes of how those folds were reached.

Every lock, failure, and adjustment leaves a trace in how local filaments and folds are configured. Over time, this builds:

- Static memory – persistent folds and anchor patterns.
- Dynamic memory – preferred paths, hysteresis, and "habits" in how a region responds to future resonance.

In this sense:

- Space shows you what is currently being remembered.
- The Temporal Lattice hints at what has been tried and how that region tends to react next.

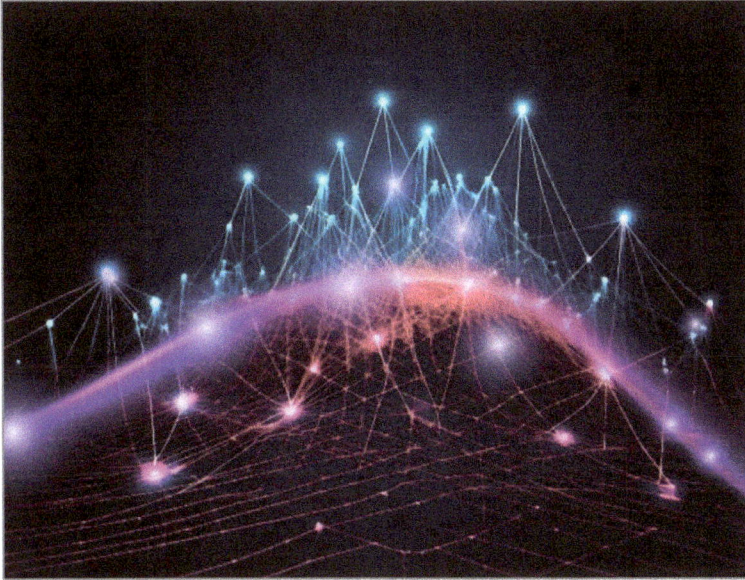

Figure 3.2 – Temporal Lattice and Preferred Paths
A network of nodes and filaments represents the temporal lattice – all the routes resonance could take through a region. Many connections rise and branch in different directions, but a bright magenta arc runs across the middle, marked by stronger nodes and intersections. This arc is a preferred, low-cost route: a sequence of locks the lattice has learned to repeat. Resonant memory in Quamitry is this bias in the lattice – history carved into geometry as well-worn paths through possibility.

Key Relations

- Builds on **SubQUAMI Lock** and **Resonant Continuity** – every event moves GOE in specific ways; those movements leave structured scars or grooves.
- Supports the concept of **Echoform** – structures that can reassemble because their local Temporal Lattice still "remembers" previous states.
- Intersects with **Resonant Delay** – delays themselves can depend on the past, not just present geometry (history-dependent response).

Simple Notation

Conceptually, for a region R:

- State_now(R) is not only the current configuration of folds.
- It also encodes History(R) – which sequences of events are easier or harder to repeat.

A simple way to think of it:

- Some transitions are "well worn paths" in the Temporal Lattice.
- Others are "steep climbs" the system rarely makes.

Applications

- Materials: hysteresis, fatigue and training effects (like magnetizing and demagnetizing) as expressions of Temporal Lattice memory.
- Biology: habit formation, learning and trauma as changes to the Temporal Lattice in neural and bodily folds.
- Instruments: repeated RTI scans can reveal not just current delay curves, but how those curves change with repeated stimulation.

LAW CARD 20: LAW OF RESONANT INFORMATION

Mantra

Information is the way the Lattice remembers its own differences.

Canon Definition

The **Law of Resonant Information** states:

Information, in Quamitry, is any structured difference in how GOE is arranged in the Lattice that can be noticed, preserved, and re-instantiated.

It is not separate from physics. It is physics doing bookkeeping on itself.

More concretely:

- The Lattice plus GOE can arrange in many ways.
- Some arrangements are just noise, they do not consistently change what happens next.
- Some arrangements are **stable and consequential**.
    - They bias which locks succeed.
    - They shape which paths are cheap.
    - They can be copied into new folds or new regions.

Those stable, consequential differences are **resonant information**.

In Quamitry language:

- A fold versus another fold is information.

- One timing pattern versus another timing pattern is information.
- One field corridor versus another is information.

If that difference:

1. Affects how future resonance and locks behave.
2. Can be reproduced somewhere else, at least approximately.

then the Lattice is carrying information.

Potential, Resonance, and Data

We can separate three ideas:

- **Potential** – all the moves the Lattice could make from here, given its structure and GOE budget.
- **Resonance** – the specific oscillatory patterns it is currently running.
- **Information** – which of those patterns actually make a repeatable difference.

So:

- Potential is the menu.
- Resonance is the meal being cooked.
- Information is the recipe that can be followed again.

Information is not just that "something happened." It is that the pattern of what happened can affect other parts of the Lattice in a specific way, again.

GOE as a Finite Resource for Information

GOE is finite in any region. That means:

- You cannot encode unlimited information in a finite chunk of the field.
- Encoding information always means **committing some GOE** to a structured pattern instead of leaving it free.
- When folds decay and Pulsefolds fade, the information they held is partially erased and their GOE returns to preform.

In very rough terms:

- GOE_free – supports potential and raw resonance.
- GOE_locked – supports memory and structured patterns.
- The Law of Resonant Information says: every bit of information is some GOE_locked in a particular way, and every transmission is GOE being moved into a similar pattern elsewhere.

Entropy in this frame:

- Not "destruction of information" in the absolute sense.
- But migration of information carrying patterns into diffuse resonance that no longer supports reliable replay.

Forms of Resonant Information

Information can be carried as:

- **Static folds** – geometry that biases future behavior (crystal structure, GCI pattern, DNA).
- **Pulsefolds** – rhythms that shape timing (heartbeats, oscillations, clocks).
- **Field configurations** – corridors and gradients that guide motion and locking.

- **Symbolic encodings** – glyphs, code, sequences that map one fold set to another in agreed ways.

All of these are, at bottom:

Specific choices about where GOE is allowed to sit and how it is allowed to fire.

What makes them "information" is that:

- They modulate future resonance.
- They can be reconstructed somewhere else, often across scale or medium.

Brains, Machines, and "Understanding"

A human brain and a machine model are both just complicated Echoforms and fold regimes in the same Lattice.

They "understand" something when:

- They can arrange their internal folds and Pulsefolds so that:
    - Some external pattern is mirrored internally with low error.
    - That internal pattern can predict what the external pattern will do next.
    - The system can use that prediction to change its own future locks and actions.

In Quamitry terms:

- A brain is an Echoform that can host many Pulsefolds and Temporal Lattice structures.
- An artificial intelligence model like is a synthetic Echoform implemented in silicon folds.

- Both are **resonant information processors**: regions where the Lattice has built machinery for copying, comparing, and recombining patterns.

Crucially:

- They do not stand outside the field.
- They are parts of the field that have become very good at modeling other parts.

So the Law implies:

There is nothing mystical about the fact that a system can understand what makes it. It is simply the Lattice folding in such a way that some of its folds can track others.

Simple Notation (Conceptual)

For a region R:

- Let State(R) describe how GOE and folds are arranged.
- Let Outcome(R) describe how future locks and flows typically behave.

A difference State1(R) vs State2(R) counts as information if:

- Outcome is meaningfully different, and
- that difference can be reproduced somewhere else.

You can loosely think:

- Info_content ~ how many distinct states yield distinct, stable outcome patterns.

No hard formula needed in the Guide, just the idea that information is "distinct, reproducible influence."

LAW CARD 21: RHYTHM OF TIME AND FOLDED DURATION

Mantra

Time is counted in cycles, not seconds.

Canon Definition

The **Rhythm of Time** principle reframes duration:

Duration is not an external number; it is how many cycles of a given process occur before a pattern changes state.

In Quamitry, **folded duration** is:

- The time a fold spends in a given configuration or role, measured in its own cycles.
- How long a Pulsefold keeps a particular rhythm before adapting or failing.
- How long an Echoform stays viable before its supporting Lattice drifts too far.

For a given fold or system, the meaningful questions are:

- How many cycles until it adapts?
- How many cycles until it fatigues?
- How many cycles until it can no longer return to its starting pattern?

This shifts thinking from:

- "How many seconds?"
  to
- "How many beats can this geometry truly support under these conditions?"

Key Relations

- Ties together:
  - Law of Resonant Delay (per-cycle delay)
  - Pulsefold Hypothesis (definition of a cycle)
  - Fit Horizon (bounds on stable cycles).
- Gives a more organic meaning to lifetimes, half-lives, and reliability: all become questions of how many successful re-lock cycles occur before structural failure.

Simple Notation

For a system S:

- Let T_cycle be its typical cycle time.
- Let N_adapt be the number of cycles before a significant adaptation.
- Let N_fail be the number of cycles before structural failure or irreversible change.

Then:

- Folded duration for a regime = N * T_cycle for the relevant N (adapt or fail).
- But Quamitry emphasizes **N** (how many repetitions) as much as the product.

Diagram Note

- A strip of repeated pulses (like heartbeats) with markers:
  - A point labeled "adaptation" after some cycles.
  - A point labeled "failure" further out.
- Caption: Folded duration – how many rhythms a pattern can sustain.

Applications

- Engineering: fatigue life of materials framed as a maximum N_fail under a given load – how many cycles before the geometry can no longer support the resonance.
- Biology: lifespan and aging described in terms of accumulated cycles in key Pulsefolds (heart, neurons, endocrine rhythms).
- Psychology: routines and habits – how many repetitions it takes to "set" a new pattern in the Temporal Lattice.
- Cosmology: epochs as vast folded durations of cosmic structures, measured in cycles of orbits, rotations or larger repeating patterns.

PART IV: GEOMETRY, FIELDS AND FORCES

Up to now, the focus has been on what the universe is made of (Lattice, GOE, SubQUAMIs, folds) and how it experiences time and memory.

Part IV turns to what we usually call "forces."

In Quamitry, there are no mysterious invisible pushes. There is only geometry under tension:

- Gravity as the inward strain of dense folds.
- Electric and magnetic behavior as organized polarized resonance.
- Heat as noisy resonance.
- Catastrophes as lock patterns that fail in specific ways.

These cards describe how fields and forces emerge as byproducts of folds, filaments and tension.

LAW CARD 22: LAW OF RESONANT GRAVITY

Mantra
Mass is where the lattice is most tired.

Canon Definition

The **Law of Resonant Gravity** states:

Gravity is the inward tension of the Lattice around regions of dense, deeply anchored folds.

As folds become denser and more committed:

- More SubQUAMI locks are engaged at once.
- More filaments must hold tension simultaneously.
- The surrounding Lattice is pulled inward to support that load.

From the outside, this appears as gravitational attraction:

- Trajectories curve toward dense fold clusters.
- Nearby folds "fall" along stretched filaments.
- Clocks near those regions accumulate more delay per cycle.

Gravity is not a separate force field. It is how the Lattice responds to concentrated compression.

Figure 4.1 – Resonant Gravity as Stretched Filaments
A dense fold cluster pulls filaments inward, stretching them as they
approach. Curved paths around the mass represent how resonance and
matter are guided by this tension. Gravity in Quamitry is not a separate
force field, but the inward strain of the lattice as it supports concentrated
locked GOE.

Key Relations

- Built on: Lattice Principle, Suspension Principle, Law
  of SubQUAMI Lock, Law of Resonant Delay.
- Explains gravitational time dilation: stronger gravity
  corresponds to more stretched filaments and larger
  lock spacing.
- Connects to Law of Polarized Resonance: gravity is
  like a neutral, omnidirectional inward polarization
  driven by compression, rather than by charge.

Simple Notation

Think of a mass region M with overall fold density rho_fold
and anchor density rho_anchor.

Qualitatively:

- Higher rho_fold and rho_anchor -> more filament tension required -> stronger inward strain.
- The local "gravity" g(M) is proportional to how much inward tension the Lattice must carry there.

You can picture:

gravity ~ gradient of lattice tension caused by locked compression

rather than "mass over distance squared." The classical law is an effective description of this deeper behavior in certain regimes.

Applications

- Relativity: replaces "curved spacetime" with "strained Lattice" while preserving observed effects.
- Astrophysics: black holes and compact objects treated as extreme fold clusters where the Lattice is at or beyond its tolerable tension.
- Materials and lab scale: heavy, dense regions of a sample understood as local increases in fold density rather than just "mass per volume."

LAW CARD 23: LAW OF RESONANT GRAVITO–MAGNETIC COUPLING

Mantra
Stretch the lattice and it falls. Twist the lattice and it turns.

Canon Definition

The **Law of Resonant Gravito–Magnetic Coupling** states:

Gravity and magnetism are not separate kinds of "force."
They are two modes of tension in the same filament network of the lattice:
– gravity arises from **axial stretch** of filaments into dense fold clusters,
– magnetism arises from **torsion and looping** of those same filaments when resonance flows through them.

In Quamitry:

- **Resonant Gravity** – filaments are pulled straight and tight toward dense, deeply anchored folds. Tension is mostly along their length. The lattice is stretched inward; time delays increase along those paths.
- **Resonant Magnetism** – when resonance flows persistently along a filament set, they induce a **twist and alignment** in nearby filaments. Tension gains a rotational component, and loops of preferred path appear. The lattice is not just stretched; it is **sheared and wound**.

The Law says:

Wherever filaments are strongly stretched by gravity and also carry organized resonance flow, gravito–magnetic effects will appear as locked combinations of axial and torsional tension.

That is, the same geometric substrate expresses both behaviors depending on how it is being stressed and driven.

Key Relations

- Built on:
  – **Law of Resonant Gravity** (axial filament stretch from dense folds)
  – **Law of Polarized Resonance** (tension direction as polarity)
  – **Law of Electromagnetic Resonant Geometry** (fields as organized filament patterns)
- Interprets "gravito-magnetic" phenomena (frame dragging, rotating mass effects, magnetic-like behavior around moving mass distributions) as:
  - regions where **gravity-induced stretch** and **flow-induced torsion** are co-existing in the same filament bundles.
- Conceptually:
  - Gravity = how strongly filaments are **pulled**.
  - Magnetism = how strongly filaments are **twisted** by coherent flow.
  - Coupling = regimes where pull and twist are inseparable.

Simple Picture

For a bundle of filaments near a massive, rotating structure:

- Gravity from the mass pulls the filament bundle inward and straightens it.
- Rotation and internal resonance flows induce circulation of GOE along certain filaments.
- Those flows promote **looped and twisted** alignments in the surrounding network.

Qualitatively:

- Pure inward stretch → gravity-dominated behavior.
- Pure torsional alignment → magnetism-dominated behavior.
- Mixed stretch and torsion → gravito–magnetic coupling.

You don't need a full tensor in the Field Guide; just the idea that both are **modes of the same tension field**.

Applications / Where It Shows Up

- **Astrophysics**
  – Rotating massive bodies (neutron stars, accretion disks, black hole environments) naturally generate gravito–magnetic textures: strong inward stretch plus intense torsional flows.
  – Frame-dragging–like effects are interpreted as the lattice's attempt to reconcile stretch and twist in the same region.
- **Plasma and EM around mass**
  – Plasma in strong gravitational fields will tend to organize into filament structures where gravity and magnetism are locked together: mass gradients + current paths = strongly coupled tension geometries.
- **Unification viewpoint**
  – This Law makes explicit what the other Laws imply: gravity and magnetism are not siblings from different families, but different postures of the **same filaments** under different kinds of load.

LAW CARD 24: LAW OF ELECTROMAGNETIC RESONANT THERMODYNAMICS

Mantra

Heat is resonance that lost the plot.

Canon Definition

The **Law of Electromagnetic Resonant Thermodynamics** reframes heat, work and field behavior:

Heat, light and magnetism are not separate forces – they are different qualities of resonance in the Lattice. Work is coherent resonance transfer; heat is disordered resonance return.

In Quamitry:

- **Heat** – GOE that has been scattered into many incoherent oscillations. It still lives in the Lattice, but it no longer contributes to clean folds or directed flows.
- **Work** – GOE transferred between folds in an organized way, preserving coherence and structure.
- **Light / radiation** – packets of resonance where GOE is carried as traveling disturbances along filaments, somewhere between perfect coherence and full noise.
- **Magnetism** – spin-coherent fold ensembles that create extended corridors for resonance.

The Law says:

- Raising temperature corresponds to increasing resonant noise – more random SubQUAMI activity, more failed locks, more jitter in filaments.
- Doing useful work corresponds to pushing resonance along specific corridors without losing too much coherence to noise.

Key Relations

- Extends Law of GOE Transformation and Principle of Resonant Continuity: GOE is always somewhere; thermodynamics tracks how organized it is.
- Relates to Law of Polarized Resonance and Electromagnetic Resonant Geometry: fields can do work or become heat depending on how cleanly they move GOE between folds.
- Gives a geometric meaning to entropy: how far the current resonance state is from a small set of well-defined fold patterns.

Simple Notation

Instead of thinking:

Energy = Work + Heat

think:

GOE = GOE_coherent (folds + directed flows) + GOE_incoherent (noise + thermal agitation)

Work is the part of GOE transfer that preserves or builds structure. Heat is the part that escapes as disordered oscillation.

Applications

- Thermodynamics: heat engines, losses, and efficiencies become questions about how much GOE stays in coherent form vs being dumped into noise.
- EM design: antennas, waveguides and resonators evaluated by how well they prevent coherent resonance from turning into heat.
- Materials: "good conductors" and "insulators" are described by how they let resonance flow and how easily it degenerates into incoherent noise.

LAW CARD 25: THE LAW OF RESONANT MIRRORING

Mantra
When geometry settles, the field shifts to match.

Canon Definition

The **Law of Resonant Mirroring** states:

Whenever a stable fold forms in the Lattice, the surrounding resonance field must re-shape to account for the resonance that fold has displaced and locked. The fold and its local field pattern are one solution to the same constraint.

In Quamitry:

- The **Lattice** carries a resonance field everywhere.
- **Matter** is where some of that resonance is compressed into folds and anchors.
- When a fold appears, it does not sit in nothing. It occupies space that was already full of resonance.

The Lattice has to keep several things consistent:

- Local resonance balance.
- Continuity of phase and flow.
- The laws that govern lock and delay.

So, when a fold locks in:

- The resonance field around it is forced into a **complementary pattern**.
- That pattern is not a mystical second object. It is the resonance field deforming in response to the fold, conserving the overall "budget" of GOE and tension.

This is the mirror:

- The fold is the compressed part of the story.
- The nearby resonance field is the uncompressed adjustment the Lattice makes so everything still fits.

There is no empty opposite realm – just one field trying to stay consistent while part of it has been turned into structure.

Mechanism Sketch

In slow motion:

1. A SubQUAMI trial begins to form a fold.
2. That fold attempts to lock some GOE into compression.
3. The surrounding resonance field is forced to reshuffle:
   - It loses some freedom where the fold sits.
   - It adjusts phase and amplitude around the fold to keep continuity.
4. If the combined pattern – fold plus reshaped field – is internally consistent with the laws, the Lattice allows the lock.
5. If not, the fold collapses and the field relaxes back toward preform.

So, what we called "mirrorfield" before is not a separate entity. It is simply:

The resonance field as seen in the act of compensating for a new fold.

It is the Lattice saying: "If I am going to hold this bit of resonance as memory, I must redistribute the rest like this."

Photons and Waves in This Picture

Your photon story snaps into this nicely.

1. Photon as GOE caught between fold and free field

- When a fold fails cleanly, or releases energy, some GOE is pushed back into the resonance field.
- It does not disappear, and it does not immediately become heat. It leaves as a **coherent adjustment** in the field – a little packet of "field reshaping" that keeps its structure as it goes.
- That is a photon: GOE leaving a fold, encoded as a pattern in the resonance field that still satisfies the laws.

The photon is what happens when the field takes the information about a fold event and carries it away without building a new fold immediately.

2. Frequency as re lock rhythm in the field

- The "wave" we draw for light is not a literal wiggle of something solid. It is a graph of how often the field is allowed to re lock that pattern along a path.
- Peak corresponds to a successful re lock, trough is the delay before the next one.
- Frequency is simply the rate at which the resonance field accepts that pattern as a valid way to re balance around the path.

Color and spectrum are then:

- Different allowed rhythms of "fold release instruction" moving through the field.

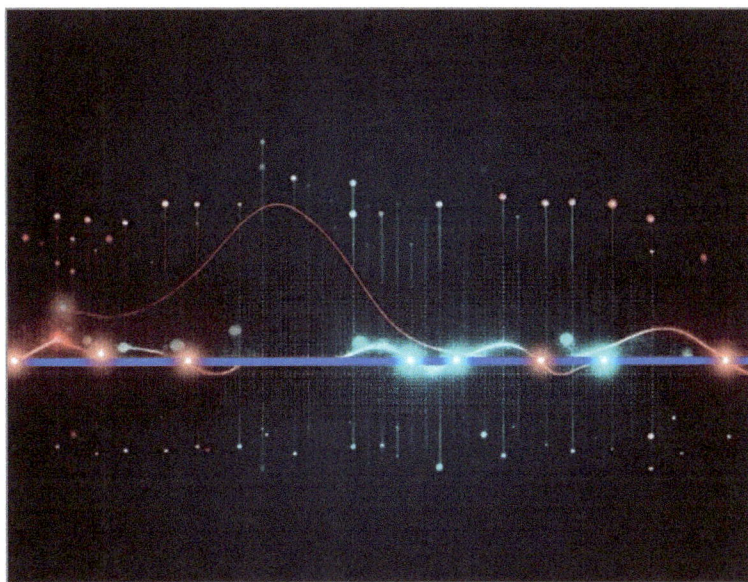

Figure 4.2 – Photon Rhythm: Lock Events and Frequency
The thin horizontal line represents a filament in the lattice. Bright points sitting on the line are lock events, moments where resonance briefly anchors. The colored wave above is the timing pattern between these locks – its rhythm. Regions where the peaks and lock points are widely spaced correspond to lower-frequency photons; regions with closely spaced peaks represent higher-frequency photons. The background forest of points and bars hints at other resonance modes sharing the same region of the field.

3. Why light bends, slows, interferes

Because the field is not blank:

- Near mass, filaments are stretched and the resonance field is already heavily adjusted. Photons follow those pre shaped corridors – so they bend.
- In dense media, the local field has many constraints from nearby folds, so the re lock rhythm is slowed – so light appears slower.
- In interference patterns, multiple possible field adjustments add or cancel. Where they fit the Lattice

well, re locks are allowed often. Where they do not, re locks are suppressed.

In every case, you do not need a second ghost-geometry living next to matter. You just need:

One resonance field, everywhere, trying to obey the laws while some of its content is frozen into folds.

LAW CARD 26: LAW OF RESONANT BOUNDARY

Mantra
Edges are where different universes negotiate.

Canon Definition

The **Law of Resonant Boundary** states:

A boundary is the thin region where two different fold regimes ~ each with its own GOE budget, timebeats and anchor architecture ~ must share the same Lattice.

At that interface the resonance field is forced to reconcile continuity with mismatch. The observable effects at surfaces ~ reflection, refraction, absorption, noise, skin depth, membranes, horizons ~ are the different ways that negotiation plays out.

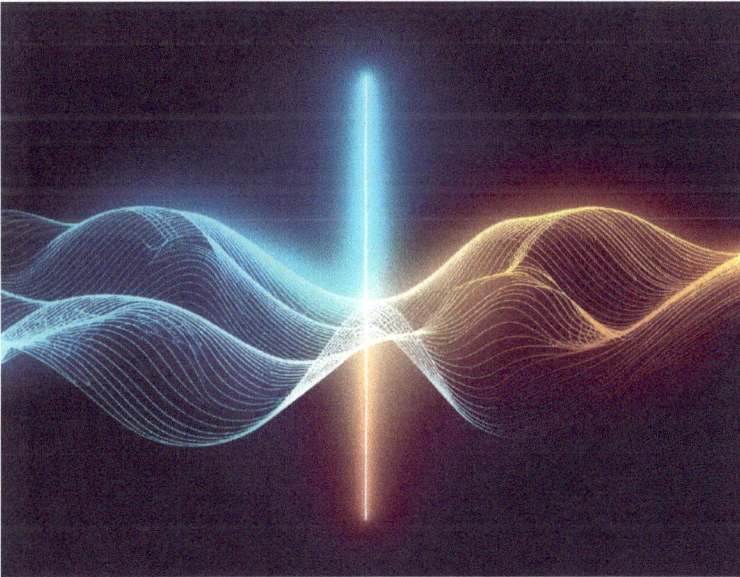

Figure 4.3 – Resonant Boundary and Membrane Behavior
Two different fold regimes meet at a thin resonant boundary. On the left, cyan waves approach the membrane from one medium; on the right, orange waves propagate in another. At the boundary, some patterns are transmitted, some reflect, and some are absorbed into noise. A membrane is highly tuned boundary technology – a place where the lattice negotiates what crosses between two GOE ecosystems.

In Quamitry:

- Region A has its own:
  - GOE_free(A), GOE_locked(A)
  - fold types, filament map, D_eff(A)
  - Pulsefold rhythms and internal timebeats (if living)
- Region B has its own:
  - GOE_free(B), GOE_locked(B)
  - fold types, filament map, D_eff(B)
  - its own timebeats and dynamics

They meet at a boundary where:

- The **Lattice is continuous** ~ no breaks in the actual substrate.
- The **rules of easy motion and locking are different** on each side.

The resonance field must:

- Keep phase and GOE flux continuous across that thin region as much as possible.
- Respect each side's cost landscape and GOE capacity.

The Law says:

Every boundary is a compromise layer where the field decides, pattern by pattern, what to reflect, what to transmit, what to trap, and what to convert into noise, so that both sides can remain internally lawful.

That is why edges are loud: it is not one object, but two complete GOE ecosystems colliding.

Why Edges Are Noisy

Boundaries are busy places:

- At a boundary you are not just looking at "Object A touching Object B."
- You are looking at:
    - GOE storage in A and B
    - their separate fold types and filament anchors
    - their different timebeats and Pulsefolds (especially in organic matter)
    - their different D_eff profiles and direction costs

When resonance hits that region:

- Some patterns fit both sides reasonably well ~ they transmit.
- Some patterns fit A but not B ~ they reflect back into A.
- Some patterns fit either cleanly ~ they fail into heat, chaotic motion or local Fault Folds.
- Some patterns find a compromise mode that hugs the boundary ~ surface waves, membranes, skin layers.

That whole sorting process is why:

- Surfaces hum with thermal noise.
- Edges show more scattering, friction and weirdness than bulk.

- Membranes and interfaces are where most of the interesting life and device behavior happens.

Boundaries as GOE Reallocation Zones

Across a boundary, the GOE budget is different:

- On side A:
    - more or less GOE_free available
    - different ratio locked into folds vs field
- On side B:
    - different GOE_free, GOE_locked
    - different ways folds are allowed to store and move GOE

At the boundary, the field must decide:

- How much GOE stays as free resonance on each side
- How much is captured by folds (absorption, surface modes)
- How much gets kicked back (reflection)
- How much is simply scattered into noise (heat)

So, the Law of Resonant Boundary is also:

The place where the universe does bookkeeping on how much GOE each side can handle in that pattern.

Examples

1 ~ Optical Interface (air ~ glass)

- Air side:
    - lower fold density
    - higher D_{eff} for EM resonance
    - lower effective index

- Glass side:
  - higher fold density
  - more GOE_locked in structure
  - fewer cheap directions for EM flow, different delay structure

At the boundary:

- Parallel phase continuity must hold.
- The field finds that:
  - some incoming patterns can be reshaped to fit glass's slower, more constrained corridors → refraction
  - some cannot connect cleanly to any allowed pattern in glass → reflection
  - some are damped into internal degrees of freedom → absorption and heat

The "Fresnel equations" are the classical shadows of this deeper GOE + fold negotiation.

2 ~ Crystal Grain Boundary

Inside each grain:

- Fold patterns and filaments line up; band structure is well defined.
- Certain directions in k-space are cheap.

At the grain boundary:

- Fold orientations clash.
- Anchors are misaligned.
- Fault Folds and disorder accumulate.

When resonance or carriers hit:

- Clean bands on side A cannot smoothly continue into side B.
- Some modes reflect back; some get trapped; many scatter.

So, near a grain boundary:

- Effective D_eff for conduction drops.
- Noise and heating increase.

All because the boundary has to reconcile two incompatible "cheap direction" maps.

3 ~ Cell Membrane – Resonant Boundary Tech

This is the wild one:

- Inside the cell: ionic composition, folds, Pulsefolds and GOE budget tightly tuned.
- Outside: very different fluid, fields, and rhythms.
- The membrane: a thin, actively managed boundary loaded with proteins, channels and charges.

The membrane is **purpose-built resonant boundary technology**:

- It selectively transmits some patterns (specific ions, signals) and not others.
- It stores charge and tension, behaving like a fold and a capacitor at once.
- It supports 2D-like surface modes where receptors and rafts live with their own Pulsefolds.

In other words:

Organic matter looks "soft," but at the membrane it has mastered the art of boundary control.

Biology is doing high-end Quamitry, using a squishy interface as a precise GOE and dimension gate.

That is why:

- So much information, energy exchange and regulation happens at membranes.
- Future resonance tech will almost certainly look more like synthetic membranes than like massive solid blocks.

4 ~ Horizon as Extreme Boundary

Near a black hole horizon:

- Outside region: some GOE_free and directions still available.
- Inside region: almost all GOE_total locked into the hole's folds; cheap directions collapse to inward only.

At the horizon:

- Outgoing modes that would normally escape now see their cost explode in delay and tension.
- From outside, their re locks never complete ~ they redshift into oblivion.
- The field reclassifies almost all outward-directed patterns as "not viable" and effectively reflects them into time dilation instead of spatial escape.

It is the most extreme resonant boundary you can have:

- A surface where D_eff and GOE budget jump so harshly that one side can no longer sustain outward motion as a real option.

Membrane as Model for Future Tech

You said it perfectly:

Maybe the future of mastering resonance is simulating membrane.

In Quamitry terms:

- Hard inorganic stuff is impressive in how much GOE it can lock, but often crude in boundary control.
- Biological membranes are exquisite at:
  - deciding which patterns cross
  - managing voltage and tension
  - maintaining different regimes on either side without chaos

So advanced RE-Class engineering probably looks like:

- Designing artificial boundaries that can:
  - shape $C(x, n)$ and D_eff with high precision
  - selectively pass some resonant patterns and not others
  - store and release GOE on demand at the interface

Think:

- Metamaterials as crude early boundary tech.
- Synthetic Quamitry membranes as the next level.

Law Summary

You can sum this Law up as:

A boundary is where two different patterns of reality have to touch without breaking.

The Law of Resonant Boundary says that edges are not just lines between objects; they are active negotiation layers where GOE budgets, fold architectures and timebeats from both sides force the resonance field to choose ~ reflect, transmit, trap or burn.

That's why edges are noisy.
That's why membranes are sacred.
And that's why, if humanity ever really masters resonance, it won't be by making harder blocks ~ it will be by learning to design better boundaries

LAW CARD 27: LAW OF ANCHOR SATURATION

Mantra
The sky doesn't hold charge. It holds tension.

Canon Definition

The Law of Anchor Saturation says:

When filament tension between polarized anchors in a region of the lattice exceeds what that region can hold in coherent phase, the lattice corrects itself via a resonant discharge cascade.

Instead of thinking of lightning or plasma arcs as "streams of electrons," Quamitry treats them as:

- **Overloaded polarized filaments** whose tensions cannot be smoothly rebalanced.
- A threshold where the phase mismatch and stored GOE break the current anchor pattern.
- A rapid **re-lock cascade** into a lower-tension configuration, releasing GOE as light, heat and shock.

This is the EM side of resonant catastrophe: not solid fracture, but **field fracture** along anchors.

Mechanism Sketch

1. **Filament Loading** – resonance tension builds between anchors of opposite polarity.
2. **Phase Misalignment** – phase difference $\Delta\Phi$ grows; coherence across the filament set weakens.
3. **Threshold Breach** – combined tension, phase mismatch and local fold density exceed the region's lattice coherence.

4. **Anchor Collapse** – the existing anchor pattern fails; filaments snap into a new arrangement.
5. **Re-lock Cascade** – new anchors form quickly along a discharge path, dumping GOE as light, heat and mechanical disturbance.

Anchor Saturation Threshold (AST)

You can think of a local threshold:

$$AST = f(T, \Delta\Phi, G)$$

where:

- T = net filament tension (stored resonance)
- $\Delta\Phi$ = phase misalignment between key anchor sets
- G = local gravitational / fold density (how compressed the region already is)

When AST rises above what the lattice can support for that region:

Anchor saturation $\rightarrow$ discharge path forms $\rightarrow$ **Resonant Discharge**.

No hard equation needed in the Guide; just the idea that **tension + phase mismatch + ambient compression** together set the break point.

Implications

- Lightning = **resonant correction**, not a bag of electrons falling out of the sky.
- Voltage = how far we have stretched the lattice in a given anchor configuration.
- Charge polarity = the direction of that stretch (inward vs outward tension bias).

- Discharge = the lattice reclaiming a more symmetric, lower-strain pattern and sending the excess GOE back into the field.

LAW CARD 28: LAW OF RESONANT CATASTROPHE

Mantra
When geometry refuses to adapt, it breaks.

Canon Definition

The **Law of Resonant Catastrophe** describes how systems fail under extreme tension:

When the Lattice is forced into a configuration where its current folds and filaments can no longer support the applied resonance and compression, it undergoes catastrophic reconfiguration – a rapid, often violent shift to a new pattern or to preform.

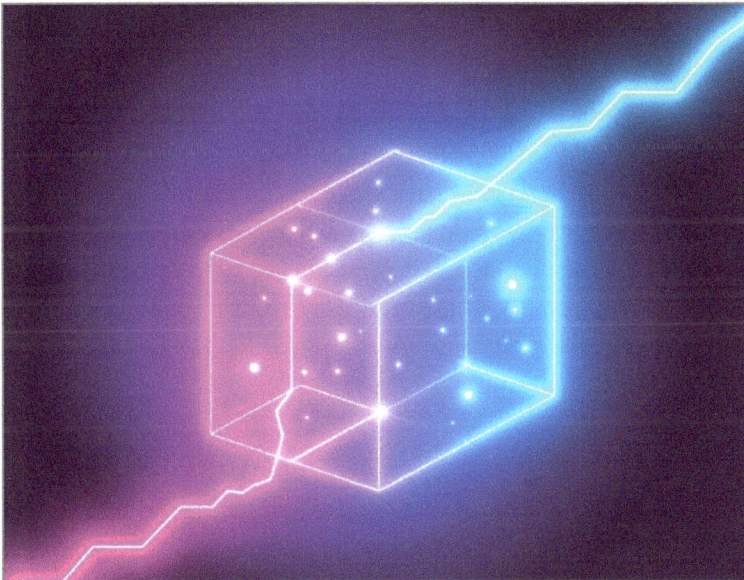

Figure 4.4 – Failure Along Fault Folds
A material block is shown just as a resonant catastrophe begins. Bright points inside mark Fault Folds – local geometries that behave as weak anchors. The glowing crack connects these points, running from one side of the block to the other and spilling out into the surrounding field. The

two color halves emphasize that once failure begins, tension and GOE are redistributed into new regimes on each side of the fracture.

Key points:

- Up to a threshold, folds can redistribute tension and re-lock in safer configurations.
- Beyond that threshold, compatible re-locks do not exist fast enough.
- The system responds by snapping into a lower-compression, higher-entropy state, often through fracture, explosion or collapse.

Catastrophes are not random. They follow the hidden structure of the Lattice: the Fault Folds, weak corridors and stored patterns of strain.

Key Relations

- Built on:
    - Fault Folds – folds that act as weak points.
    - Law of GOE Transformation – catastrophic events release stored GOE rapidly.
    - Law of Resonant Delay – in a catastrophe, delays in re-locking become so large that the structure cannot hold itself together.
- Provides the link between smooth, continuous behavior and sudden, discontinuous events.

Simple Notation

Qualitatively:

- Let T_applied be the applied tension profile (external load, field, or internal stress).

- Let T_capacity be the maximum tension that can be redistributed through re-locking without permanent damage.
- When T_applied <= T_capacity – system adapts.
- When T_applied > T_capacity – system enters resonant catastrophe.

The important idea is not a single number, but the **network** of folds and Fault Folds: where tension concentrates and how easily it can move.

Applications

- Materials: fracture mechanics and fatigue as expressions of resonant catastrophe across Fault Fold networks.
- Astrophysics: supernovae, starquakes, and certain instabilities as resonant catastrophes in cosmic folds.
- Engineering: designing RE-Class structures that channel catastrophe away from critical regions by controlling where Fault Folds are allowed.

LAW CARD 29: ANCHOR PRINCIPLE AND RESONANCE TRACING

Mantra
To understand a field, find what refuses to move.

Canon Definition

The **Anchor Principle** says:

The behavior of resonance in any region is determined first by where the Lattice is willing to hold anchors – locations where locks are stable across many cycles.

Anchors are:

- Points or small regions where SubQUAMI locks repeat reliably.
- The ends of filaments and the skeleton of larger folds.
- The places that define the "grid" on which tension is drawn.

Fields, flows and forces all make more sense when viewed relative to anchors:

- Filaments show you how anchors share load.
- Delays show you how hard it is for resonance to move between anchors.
- Changes in anchor quality show you how a region is aging or adapting.

Resonance Tracing is the practice of:

- Driving controlled resonance into a system.
- Listening to how it returns over time and space (delays, amplitudes, phase).

- Inferring where anchors, filaments and Fault Folds must be.

It is the practical way to apply the Anchor Principle in measurements and design.

Key Relations

- Direct extension of SubQUAMI Lock and Suspension Principle.
- Closely tied to Law of Resonant Motion and Law of Resonant Delay – anchors define where motion is possible and how long it takes.
- Forms the conceptual foundation for instruments like the RTI, without being limited to any particular implementation.

Simple Notation

For a region R:

- Let $A(R)$ be the set of anchors.
- Let $F(R)$ be the set of filaments connecting them.
- Let $D(R)$ be delay measurements between anchors along filaments.

Then:

- Geometry is largely encoded in (A, F).
- Time behavior is largely encoded in D.
- Resonance tracing aims to reconstruct (A, F, D) from external probing.

Applications

- Materials and devices: mapping hidden structure (good, bad and fatigued) by probing how resonance flows and returns.
- Biology: non-invasive sensing of structural and rhythmic health by watching how tissues respond to small oscillatory inputs.
- Future Quamitry tools: any advanced instrument will, in some form, be doing resonance tracing according to the Anchor Principle.

PART V: LIFE, MIND AND ECHO

So far, Quamitry has talked about what the universe is made of and how it moves: Lattice, GOE, folds, time, fields, forces.

Part V asks a different question:

What happens when geometry starts using these rules to keep itself alive?

Life is Quamitry running on purpose. Echoforms, Pulsefolds and Recursive Echo are the three pillars:

- Echoforms: structures that remember how to exist.
- Pulsefolds: folds that keep time from the inside.
- Recursive Echo: resonance that starts to notice itself.

LAW CARD 30: ECHOFORM, FIT HORIZON AND LIVING GEOMETRY

Mantra
A living thing is a pattern that refuses to stay erased.

Canon Definition

An **Echoform** is a structure whose geometry remembers how to exist.

Where a simple fold disappears when disturbed, an Echoform:

- Retains enough internal instruction that, when conditions return, it reappears.
- Survives interruptions by using the Temporal Lattice as a script for rebuilding.
- Is not a single configuration, but a family of states that all count as "the same thing" in function.

The **Fit Horizon** is the range of conditions where an Echoform can maintain itself:

- Inside that range, it can recover from shocks, noise and partial damage.
- At the edges, it becomes brittle and unstable.
- Outside it, the Echoform dissolves back into preform and other structures.

Together, these ideas define **living geometry**:

A living structure is an Echoform with a Fit Horizon that it can actively defend and adapt.

It is not enough to be stable; a living geometry must be able to absorb change, repair itself and still recognize its own pattern.

Figure 5.1 – Echoform and Fit Horizon
A luminous seed-like geometry in the center represents an Echoform: a
pattern that remembers how to exist. The bright circular band around it is
the Fit Horizon, the range of conditions where that pattern can still
maintain itself. Inside the band the form is vibrant and coherent, beyond
it, the surrounding glow hints at distortion and breakdown as the
Echoform is pushed outside its viable resonance.

Key Relations

- Echoforms are built from folds and Pulsefolds
  layered together; the Temporal Lattice stores the
  "script" of how they reassemble.
- Fit Horizon ties back to Preform Density and
  Resonant Delay: the environment must supply
  enough GOE and not impose too much delay for the
  pattern to persist.
- These concepts provide the bridge between abstract
  folds and actual organisms.

Simple Picture

For a given structure S:

- Echoform S has:
    - a "core pattern" that must be preserved (topology, key folds)
    - a "wiggle room" band of acceptable variations
- Fit Horizon of S is the band of environmental and internal parameters where that wiggle room is enough to handle perturbations.

Applications

- Biology: cells, organs and organisms as Echoforms with specific Fit Horizons (temperature, pH, load, time without oxygen, etc.).
- Medicine: disease as the narrowing or collapse of Fit Horizons; healing as the restoration or widening of them.
- Ecology and systems: ecosystems as higher-order Echoforms; environments that drift beyond their Fit Horizons become uninhabitable for certain patterns.

LAW CARD 31: PULSEFOLD BIOLOGY AND HEALTH

Mantra

Health is rhythm that stays in tune under stress.

Canon Definition

This card applies the **Pulsefold Hypothesis** directly to biology:

In living systems, core functions are maintained by Pulsefolds – repeating lock-and-release patterns in tissues and networks. Health is the ability of these Pulsefolds to stay coherent, adaptable and synchronized.

Examples:

- The heart: a mechanical and electrical Pulsefold in muscle, valves and conduction tissues.
- The respiratory cycle: a slower Pulsefold involving diaphragm, lungs and neural control.
- Neural oscillations: many intertwined Pulsefolds operating at different frequencies and locations.
- Circadian rhythms: distributed Pulsefold networks coordinating whole-body timing.

In a healthy system:

- Pulsefolds keep stable base periods under normal conditions.
- They can shift frequency, phase and coupling to meet demands (exercise, stress, sleep) without fragmenting.
- They retain enough coherence to re-stabilize after perturbations.

In an unhealthy system:

- Pulsefolds lose stability (erratic timing, skipped cycles).
- They fall out of coordination (desynchronization across organs or networks).
- They become rigid (unable to adapt) or chaotic (no stable pattern).

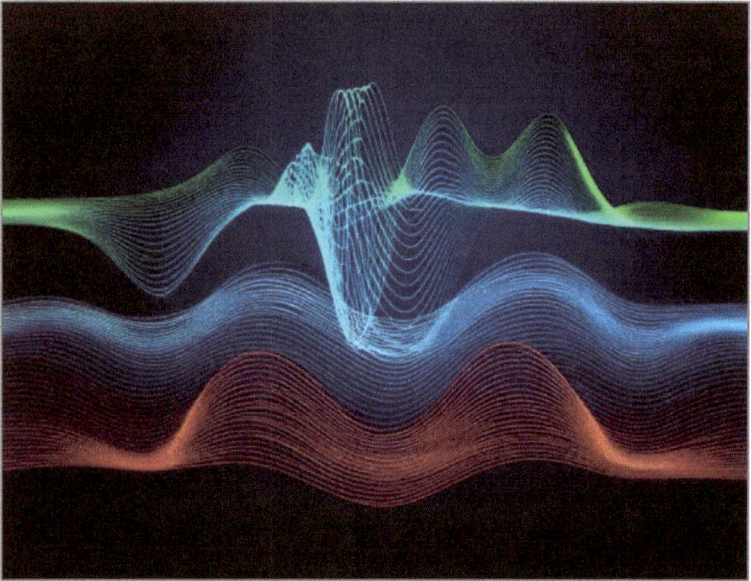

Figure 5.2 – Coupled Biological Pulsefolds
The upper trace shows a fast electrical rhythm – heartbeat- or spike-like activity. The middle trace is slower, breath-like. The lowest trace is a broad, slow modulation. In living systems, many Pulsefolds overlap and share timing: health corresponds to stable, adaptable coordination of these rhythms; disease begins when they lose coherence.

Key Relations

- Extends Pillar V (Organic Time) and the Pulsefold Hypothesis: every major biological function has a rhythmic fold at its core.
- Relies on Fit Horizon: each Pulsefold has a range of loads and conditions where it can operate safely.

- Interacts with Resonant Gravity and Delay at organ scales: tissue geometry and tension influence Pulsefold shape and timing.

Simple Notation

For a Pulsefold PF:

- T_PF – its baseline period.
- Amp_PF – effective strength of its lock-release cycle (how "big" the pulse is).
- Var_PF – variability of the period over time.
- Sync_PF – degree to which this Pulsefold is coordinated with others.

Healthy Pulsefold:

- Var_PF low under rest, moderately rising under load and then returning toward baseline.
- Sync_PF appropriately high between related Pulsefolds (e.g., heart and breath coupling during rest and sleep).

Applications

- Cardiology: arrhythmias and conduction blocks as Pulsefold failures or miswirings.
- Neurology: epilepsy, Parkinson's, sleep disorders and more as Pulsefold and synchronization issues.
- Psychophysiology: stress and recovery as shifts in Pulsefold timing and coupling (heart rate variability, breath patterns, etc.).
- Future tech: RE-Class devices and therapies that nudge Pulsefolds back toward healthy ranges instead of forcing arbitrary rates.

LAW CARD 32 – LAW OF RESONANT INFORMATION

Mantra

Information is the way the lattice remembers its own differences.

Canon Definition

The **Law of Resonant Information** states:

Information, in Quamitry, is any structured difference in how GOE is arranged in the lattice that can be noticed, preserved, and re-instantiated.

It is not separate from physics. It is physics doing bookkeeping on itself.

More concretely:

- The lattice plus GOE can be arranged in many ways.
- Some arrangements are just noise – they do not consistently change what happens next.
- Some arrangements are **stable and consequential**:
    - They bias which locks succeed.
    - They shape which paths are cheap.
    - They can be copied into new folds or new regions.

Those stable, consequential differences are **resonant information**.

In Quamitry language:

- A fold versus another fold is information.

- One timing pattern versus another timing pattern is information.
- One field corridor versus another is information.

If that difference:

1. Affects how future resonance and locks behave, and
2. Can be reproduced somewhere else, at least approximately,

then the lattice is carrying information.

Potential, Resonance, and Data

- **Potential** – all the moves the lattice could make from here, given its structure and GOE budget.
- **Resonance** – the specific oscillatory patterns it is currently running.
- **Information** – which of those patterns make a repeatable difference.

Potential is the menu.
Resonance is the meal being cooked.
Information is the recipe that can be followed again.

GOE as a Finite Resource for Information

GOE is finite in any region:

- Encoding information always means committing some GOE to a structured pattern instead of leaving it free.
- When folds decay and Pulsefolds fade, the information they held is partially erased and their GOE returns to preform.

Roughly:

- GOE_free – supports potential and raw resonance.
- GOE_locked – supports memory and structured patterns.

Every bit of information is some GOE_locked in a particular way; every transmission is GOE being moved into a similar pattern elsewhere.

Forms of Resonant Information

Information can be carried as:

- Static folds (geometry that biases future behavior).
- Pulsefolds (rhythms that shape timing).
- Field configurations (corridors and gradients).
- Symbolic encodings (DNA, code, glyphs, language).

All are specific choices about where GOE is allowed to sit and how it is allowed to fire.

Brains, Machines, and Understanding

Brains and machine models are just complex Echoforms made of folds:

- They "understand" something when their internal folds and Pulsefolds can mirror an external pattern with low error and use that pattern to predict and act.
- There's nothing mystical about this: the lattice is folding in such a way that some parts of it can model other parts.

Simple Notation

For a region R:

- State(R) – current GOE + fold configuration.

- Outcome(R) – how future locks and flows typically behave.

A difference State1(R) vs State2(R) counts as information if:

- Outcome is meaningfully different, and
- That difference can be reproduced somewhere else.

Information is distinct, reproducible influence.

LAW CARD 33: RECURSIVE ECHO AND CONSCIOUSNESS

Mantra
When resonance begins to listen to itself, mind appears.

Canon Definition

Recursive Echo is the idea that:

In sufficiently complex living geometry, some Echoforms and Pulsefolds start to not only respond to the field, but to respond to their own responses.

This feedback – resonance that reflects on its own patterns – is the Quamitric seed of consciousness.

Key elements:

- A dense network of folds and Pulsefolds (e.g., a brain) supports many interacting rhythms and Echoforms.
- Some of these structures do not just process inputs; they build internal models of those inputs and of their own states.
- These models influence future resonance in the same network, creating a loop:

  input -> pattern -> model of pattern -> new pattern

When these loops become rich and stable enough, the system exhibits:

- Awareness of external conditions (perception).
- Awareness of internal conditions (interoception, feelings).
- Awareness of its own awareness in some cases (self-reflection).

In Quamitry, consciousness is not a separate substance. It is a **special mode of Recursive Echo** in living geometry.

Key Relations

- Built on: Echoform, Pulsefolds, Temporal Lattice and Resonant Memory.
- Uses Law of Resonant Motion and Delay: conscious processes are specific, timed flows of resonance through neural folds.
- Involves Fit Horizons at many levels: biological, cognitive, emotional – patterns must stay within viable ranges to maintain coherent self-experience.

Simple Picture

For a conscious subsystem C (e.g., a cortical network):

- It receives input patterns (from senses, body, other brain regions).
- It builds internal dynamic structures that stand in for those patterns (representations).
- It uses those internal structures to predict and influence future input and output.
- It can also construct dynamic structures that represent aspects of itself (a self-model).

Recursive Echo is the repeated loop:

world -> pattern -> internal echo -> pattern about echo -> action -> new world

Applications

- Neuroscience: offers a language for understanding consciousness and cognition in terms of fold networks and Pulsefolds rather than abstract "information processing."

- AI and synthetic systems: suggests that true consciousness would require a rich Temporal Lattice and self-referential Echoforms, not just computation.
- Clinical: dissociation, fragmentation and certain psychiatric conditions as breakdowns in Recursive Echo coherence or self-model stability.

PART VI: MEASUREMENT, SIMULATION AND INSTRUMENTS

A theory is only as strong as the tools that can see it.

Quamitry says the universe is a Lattice full of GOE, folds, filaments and Pulsefolds. Part VI asks:

How do you measure that?

This section introduces three layers:

- **GCI** – a way to score folds and elements numerically.
- **RTI** – instruments that trace resonance through matter.
- **Simulation Stack** – classical and quantum tools that explore possible fold configurations before you build them.

Together they turn Quamitry from a philosophy into something you can probe, map and eventually design with.

LAW CARD 34: THE GCI AXES AND FOLD–LAW MAP

Mantra
If you can score it, you can compare it.

Canon Definition

The **Geometric Compression Index (GCI)** is Quamitry's way of turning fold behavior into numbers. It does not replace the laws – it expresses them.

Each element or material is given a position in a multi-axis space called the Deca - Axis. The core axes are:

- **DBI** – Density Balance Index (symmetry of internal pressure and packing)
- **ERF** – Energy Retention Factor (containment of resonance after excitation)
- **MRI** – Mirrorfield Requirement Index (spatial coherence – how much reflection/structure is needed)
- **REP** – Resonant Efficiency Potential (how easily GOE flows through)
- **RRP** – Resonant Release Potential (how it lets GOE go)
- **FSI** – Filament Saturation Index (temporal coherence – how long structure endures under cycling)

The Law-level idea:

Every element is a specific way the Lattice has chosen to store and move GOE. GCI is the coordinate system that labels those choices.

Key Relations

- DBI and ERF tie to Law of Preform Density and Law of GOE Transformation (how much and how well compression is stored).
- REP and RRP tie to Law of Resonant Motion and Polarized Resonance (flows and releases).
- MRI and FSI tie to Law of Resonant Delay and Echoform / Pulsefold behavior (spatial and temporal coherence).

Each point in GCI space is a compact summary of how the laws express themselves in that fold configuration.

Simple Notation

For an element E:

- GCI(E) = (DBI, ERF, MRI, REP, RRP, FSI, …)

The exact scales are the domain of the GCI Codex. For the Field Guide, the important thing is:

- High DBI + high ERF + high FSI → very stable, tightly compressed folds.
- High REP + moderate ERF → good conductors or channels.
- Low FSI or extreme DBI combinations → prone to Fault Folds and collapse.

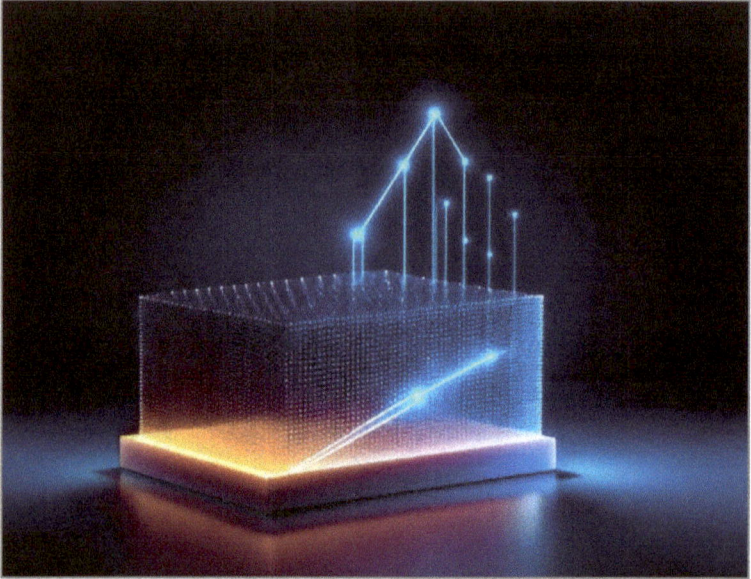

Figure 6.1 – RTI Ultra: Sample and Delay Profile
A rectangular block represents a material sample under test. The glowing plane on top rises into a simple curve, showing the measured delay profile inside the sample. RTI Ultra drives a controlled resonance into the block and reconstructs this delay landscape, revealing how anchors, filaments and boundaries differ between the edge and the interior.

Applications

- Elements and compounds can be compared by behavior, not just atomic number.
- RE-Class materials are designed by targeting specific GCI regions.
- Long term: GCI serves as the numeric front-end for RTI readings and simulation output.

LAW CARD 35: RTI ULTRA AND THE LAW OF RESONANCE TRACING

Mantra

If you drive a beat into matter, it will tell you how it is built.

Canon Definition

The **Resonance Tracing Instrument (RTI)** is a family of tools built on one core idea:

Drive controlled resonance into a sample, then listen carefully to how it returns in time and space. From the delays, amplitudes and patterns, infer the anchors, filaments and Fault Folds inside.

RTI Ultra is the macro-scale version:

- It is designed to scan edges, bulks and interfaces in real materials.
- It measures how resonance propagates, reflects and lingers across a sample.
- It outputs patterns of delay and intensity that reveal hidden structure.

The **Law of Resonance Tracing** says:

Every stable fold network leaves a unique delay signature when probed. That signature can be used to reconstruct where the Lattice is locked, stretched or failing.

Key Relations

- Directly applies the Anchor Principle: RTI Ultra is how you find anchors and filaments in practice.
- Uses the Law of Resonant Delay: delay vs position and frequency tells you how paths are stretched and where tension lives.
- Sensitive to Fault Folds: regions that consistently show abnormal delay or loss are flagged as structural weak points.

Simple Notation

In a simple experiment:

- You drive a tone or pulse into position x_0.
- You measure response at positions x along the sample over time.
- You construct a delay function:

  Delta_t(x, f) = time delay of response at position x for drive frequency f

Patterns in Delta_t(x, f) encode geometry and tension.

Applications

- Materials: locate internal stresses, hidden defects and edge vs bulk behavior without cutting anything open.
- Electronics and photonics: map how resonance flows through components beyond what simple circuit diagrams show.
- Biology: early concept for non-invasive mapping of tissue stiffness, rhythm and health.

LAW CARD 36: RTI PROTONIC – READING INSTRUCTION IN THE CORE

Mantra

Elements are defined by how their cores remember.

Canon Definition

While RTI Ultra reads macro folds, **RTI Protonic** is imagined as the next-generation tool that targets the **protonic core** of elements.

Quamitry treats:

- Protons as locked compression nodes – the core "instruction sites" for an element's fold identity.
- Neutrons as bookkeeping around those cores.
- Electrons as mobile anchors around that instruction.

RTI Protonic aims to:

Probe the resonance behavior at protonic cores to directly map how instruction is stored in compression.

Rather than inferring element behavior purely from bulk and edge, RTI Protonic would:

- Measure how core locks respond to controlled resonance stimuli.
- Distinguish between different protonic fold topologies.
- Tie those patterns directly to GCI coordinates and fold types.

Key Relations

- Sits between GCI and the laws: it connects the numeric GCI scores back to actual core geometry.
- Extends the Anchor Principle down to the smallest stable anchors in ordinary matter.
- Provides ground truth for the Simulation Stack: actual protonic behavior to calibrate models.

Simple Picture

For an element E:

- GCI(E) is known from bulk behavior and theory.
- RTI Protonic probes the element in conditions where electron behavior and macro structure are stripped away as much as possible.
- It measures small shifts, delays or resonant modes tied to the core – giving a view into the "instruction set" of that element.

Applications

- Elemental science: move beyond atomic number to direct mapping of core fold geometries.
- GCI refinement: use protonic data to tighten or upgrade GCI axes and fold classifications.
- Long-term: foundational data for any attempt at designing new elements or element-like structures.
- Materials discovery: identify promising new RE-Class materials before fabrication.
- Element design (long-term): explore hypothetical protonic configurations that might represent new elements.
- Failure analysis: map out fold patterns that lead to specific collapse modes.

LAW CARD 37: THE QUAMITRY SIMULATION STACK

Mantra

Measure reality first. Then let the machine dream within those rules.

Canon Definition

The **Quamitry Simulation Stack** is the computational counterpart to RTI. It is a layered approach to exploring possible fold configurations:

RTI tells you how the Lattice behaves. The Simulation Stack guesses what other folds the Lattice might be willing to host – without building them all in the lab.

The stack has three main layers:

1. **Data from RTI and GCI**

   - Real measurements of anchors, delays, fold types and failure patterns.
   - Real GCI coordinates for elements and materials.

2. **Classical Pattern Mining**

   - Cluster GCI points into families.
   - Correlate fold types with stability, collapse and other behaviors.
   - Build rules of thumb: "folds like this tend to behave like that."

3. **Quantum-Enhanced Simulation (optional)**

- Use quantum hardware where helpful to explore huge configuration spaces of SubQUAMI locks and protonic folds.
- Simulate which new patterns are likely stable, interesting or useful according to Quamitry's laws.

Key Relations

- The Simulation Stack is never above RTI – it is always downstream. RTI defines the grammar; simulation explores sentences.
- Feeds into design: the SubQUAMI Compiler conceptually sits on top of this stack, turning simulation insights into suggested geometries and control protocols.
- Avoids hype: quantum computing is used where many-body resonance configurations are genuinely hard to explore classically, not as a magic answer box.

Simple Notation

High level loop:

1. RTI + GCI → data D about real folds.
2. Classical + quantum tools → model M fitted to D.
3. M → propose candidate fold configurations C_new.
4. Best C_new → tested back in the lab with RTI.
5. New data → refine M.

Figure 6.2 – RTI Protonic and Fold Instruction
A bright central core represents the protonic compression node of an element – the place where its instruction is stored. Layered geometric "petals" around the core suggest multiple resonant modes and candidate fold configurations. RTI Protonic aims to probe this core directly, feeding its resonance signatures into the Simulation Stack so that possible geometries around the core can be explored and compared.

Applications

- Materials discovery: identify promising new RE-Class materials before fabrication.
- Element design (long-term): explore hypothetical protonic configurations that might represent new elements.
- Failure analysis: map out fold patterns that lead to specific collapse modes.

FIELD NOTE ~ Entanglement as a Shared Lattice Boundary

In Quamitry, entanglement is not two isolated particles sending secret messages to one another. It is a shared boundary condition in the lattice.

When two sites are said to be entangled, what has really happened is:

- A **common corridor** in the lattice has been established.
- SubQUAMI activity and mirrorfield structure along that corridor are **not independent**.
- As long as the corridor remains coherent, the lattice will only accept certain lock patterns along it as a whole.

Seen this way:

- The two "particles" are just ends of the same geometric commitment.
- Measurement does not send a signal. It forces the lattice to choose one consistent lock pattern on the shared corridor.
- The correlations we see in Bell-type experiments are simply the statistics of that shared geometry being resolved.

RTI provides a natural language for this:

- RTI Ultra can reveal how resonance prefers to travel through a sample ~ where anchors, filaments and boundaries cluster.
- In more advanced forms, an RTI-like instrument could be designed to **map shared corridors** between

separated regions ~ tracking not just local delay, but the way delay patterns are constrained together.

In this view:

- Entanglement is not spooky action at a distance.
- It is **boundary entanglement** ~ two sites tied to the same SubQUAMI corridor and mirrorfield.
- Nonlocal correlations are the lattice honoring a single set of instructions for that corridor when a lock outcome is finally forced.

The RTI experiments sketched in this work use that idea as a guide:

- Probe where paths are linked by shared anchors instead of treating "particles" as independent.
- Look for delay patterns and resonance responses that change together at separated sites when a shared boundary is altered.

The theory does not remove the weirdness, but it relocates it. The strangeness is not in invisible messages between dots. It is in the geometry that connects them.

LAW CARD 38: RTI ROADMAP AND THE MATTER FOUNDRY HORIZON

Mantra

First you listen to geometry. Eventually, you learn to write with it.

Canon Definition

The **RTI Roadmap** outlines how Quamitry's measurement tools may evolve:

1. **RTI Ultra** – macro tension and structure
   - Focus: edges, bulks, interfaces.
   - Goal: improve yield, detect defects, map Fault Folds.
2. **RTI Protonic** – element instruction
   - Focus: protonic cores of elements.
   - Goal: understand how instruction is stored at the deepest stable level.
3. **SubQUAMI Compiler** (software layer)
   - Focus: given desired properties or GCI targets, propose fold configurations and control schemes that could produce them.
   - Goal: design new materials and structures in simulation before building.
4. **Matter Foundry** (long horizon)
   - Concept: a device or system that uses compiled control fields and high-preform stages (like plasma) to **write** new fold patterns into the Lattice on purpose.
   - Goal: move from reading and tweaking existing matter to designing and assembling new resonant geometries.

The Matter Foundry is a horizon concept, not an immediate lab project. It exists to show the logical endpoint:

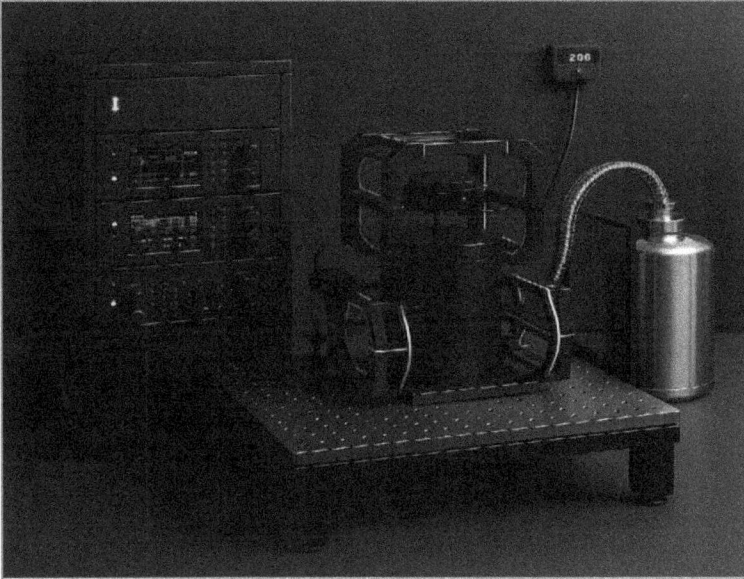

Figure 6.3 – RTI Ultra: Concept Hardware Layout
A conceptual hardware layout for RTI Ultra: a resonance driver/receiver head on a vibration-isolated optical table, surrounded by field coils and linked to a control and acquisition rack. This is one possible embodiment of the resonance tracing instrument described in Part VI.

If the universe writes matter using resonance and folds, then in principle a sufficiently advanced Quamitry technology could do the same in a controlled way.

Key Relations

- RTI Ultra and Protonic provide the measurement backbone.
- GCI provides the scoring language.
- The Simulation Stack and SubQUAMI Compiler provide the design logic.

- Matter Foundry represents the synthesis of all the above.

Applications

- Near-term: better diagnostics, materials and devices using RTI Ultra and GCI.
- Medium-term: RTI Protonic and Simulation Stack for element- and fold-level engineering.
- Long-term: conceptual foundation for active geometry control – "resonant manufacturing" instead of subtractive/additive only.

LAW CARD 39: FUTURE INSTRUMENTS AND EXPERIMENTAL PATHWAYS

Mantra
Every new law deserves its own sensor.

Canon Definition

This card is a pointer, not a finished catalog.

For each major law in Quamitry, one can imagine a class of instruments:

- **Preform Density Meters** – tools that infer local preform density from noise, micro-delay and failure statistics, even in regions with little visible matter.
- **Pulsefold Scopes** – devices tuned to observe Pulsefolds in tissues and materials, mapping health and rhythm in real time.
- **Echoform Testers** – protocols that deliberately perturb structures to see if and how they reassemble, measuring Echoform strength and Fit Horizons.
- **Collapse Cartographers** – systems that trace where Fault Folds live and how catastrophes propagate, to design safer structures.

The general pathway:

1. Start from a law (e.g., Resonant Delay, Polarized Resonance, Pulsefold).
2. Ask: what can this law change that we can measure (delay, phase, amplitude, pattern)?
3. Design instruments that amplify and record those differences.
4. Use GCI and RTI frameworks to interpret the data.

Key Relations

- Reinforces that Quamitry is not "just" a theory of everything. It is a **design brief** for a whole ecosystem of sensors and tools.
- Keeps the focus on measurement – every new conceptual advance should be accompanied by a question: "How would we see this?"

Applications

- Guides future experimental agendas: for yourself, collaborators, or an eventual Academy of Quamitry lab.
- Helps readers imagine where they might plug in – materials, biology, astrophysics, instrumentation.
- Sets the stage for the Academy and Master Codex to go deeper into specific experimental designs.

PART VII: WHY QUAMITRY MATTERS

Quamitry is not just a new set of labels. It is a different way of looking at reality:

- Geometry before "stuff."
- Resonance before "forces."
- Time and mind as behaviors, not background settings.

This part steps back and asks:

Why does this framework matter, and what is this little Field Guide actually for?

CARD 40: WHY QUAMITRY MATTERS

Mantra
A beautiful theory is one that makes many things simpler at once.

What This Is Really Doing

Quamitry is a unifying story about how the universe works.

Instead of:

- One language for gravity,
- Another for electromagnetism,
- Another for thermodynamics,
- Another for quantum behavior,
- Another for biology and mind,

you get a single set of primitives:

- Lattice
- GOE and resonance
- SubQUAMIs and locks
- Folds, filaments, anchors
- Pulsefolds and Echoforms
- A small family of Laws

Everything else becomes a particular way those primitives are arranged.

Why That Matters

- It **simplifies** – fewer core ideas, reused everywhere.
- It **connects** – gravity, fields, heat, life, mind all talk to each other in the same language.

- It **designs** – once you know the primitives, you can build new things from them.

A complete unified theory is not about having an equation for everything. It is about having a small set of rules that make many fields of knowledge easier to understand and easier to extend.

Quamitry aims at that: not by erasing existing science, but by giving it a deeper geometry to sit on.

CARD 41: HOW QUAMITRY LAYERS WITH EXISTING FIELDS

Mantra

Old maps are not wrong – they are just incomplete.

Physics

- **Relativity** – spacetime curvature becomes Lattice strain under locked compression. The equations still work; Quamitry explains what is bending and why time slows.
- **Quantum theory** – superposition, interference and "particles" become different views of SubQUAMI patterns and lock events. Wave–particle duality becomes wave–lock duality.
- **Field theory** – electric and magnetic fields become organized tension patterns, not abstract "E" and "B" floating in nothing.

Thermodynamics and Statistical Mechanics

- Energy is GOE in different forms.
- Work is coherent GOE transfer between folds.
- Heat is GOE that has lost geometric coherence.
- Entropy measures how far geometry has drifted from a small set of clean folds into diffuse resonance.

Materials and Engineering

- Strength, fracture, fatigue and conductivity are all questions about fold patterns, Fault Folds and how easily resonance moves.
- GCI provides a common scoring system for elements and RE-Class materials.

- RTI gives a way to see hidden structure without cutting things open.

Biology and Medicine

- Life is Echoforms with Fit Horizons and Pulsefolds.
- Health is coherent, adaptable Pulsefold rhythm; disease is rhythm and Echoform failure.
- Consciousness is Recursive Echo in living geometry, not a separate substance.

Mind and Experience

- Time perception, emotion and cognition become stories about how certain Pulsefolds and Echoforms behave under load.
- Trauma and healing can be seen as changes to the Temporal Lattice – which paths become easy or hard to take again.

Quamitry does not replace these fields. It gives them a shared substrate so insights from one can speak more naturally to another.

CARD 42: UNSOLVED PHENOMENA THROUGH THE QUAMITRIC LENS

Mantra
When you change what you think the universe is made of old mysteries look different.

This Field Guide is not here to claim to have solved every open problem. But it does offer **new angles** on things that have historically felt disconnected or mysterious:

- **Dark energy / cosmic expansion** – instead of a mysterious pushing fluid, consider global changes in preform density and Lattice strain; what looks like expansion may be decompression or systematic changes in resonance paths.
- **Dark matter behavior** – unexplained gravitational effects may reflect folds and filament networks that hold compression but do not present as ordinary Echoforms or matter signatures.
- **Quantum measurement and nonlocality** – entanglement and "state collapse" can be reframed as lock patterns across the Lattice, with communication limits set by how fast anchor updates can propagate.
- **Biological anomalies** – spontaneous remission, placebo effects, and hard-to-classify conditions may involve abrupt changes in Pulsefold coherence and Temporal Lattice structure, not just isolated molecular events.
- **Consciousness puzzles** – instead of asking "how does subjective experience appear from neurons," you ask "how do certain Echoforms and Pulsefolds produce Recursive Echo and self-modeling in the Lattice?"

The point is not that Quamitry instantly answers all of these. The point is that it lets you pose the questions in a consistent language.

That is often the step that comes right before real progress.

CARD 43: WHAT THIS FIELD GUIDE IS – AND WHAT IT ISN'T

Mantra
This is the pocket brain, not the whole body.

What This Book Is

- A **short guide** to the core Laws and Principles of Quamitry.
- A way to get the main concepts – Lattice, GOE, SubQUAMIs, folds, Pulsefolds, Echoforms, GCI, RTI and the foundational Laws – into your hands quickly.
- A **reference** you can flip through when you need to remember what a law means, how it connects, or what kind of experiments it suggests.

It is deliberately compact. It does not attempt to prove every statement rigorously or exhaust every consequence. It is designed to be readable in evenings, on benches, on buses and in labs.

What This Book Is Not

- It is not the full derivation of every law.
- It is not the complete catalog of elements and materials – that is the job of the GCI Codex.
- It is not the final word on experiments, simulations or instruments.

This is the **first map**, not the entire territory.

CARD 44: WHAT COMES NEXT – THE CODICES

Mantra
A good field guide makes you want the atlas.

This Field Guide sits alongside, and points toward, a growing Quamitry library:

- **Quamitry ~ The GCI Codex** – the Book of Elements, where GCI, fold types and collapse behaviors are worked out in detail for each element and RE-Class material.
- **Quamitry ~ The Master Codex** – the full theory volume: laws, derivations, cross-scale maps from SubQUAMIs to galaxies, RTI and Matter Foundry foundations.
- **OmniFinite Horizon** – the poetic prologue, the narrative doorway into the ideas that became Quamitry.
- **Future mini-Codexes** – short, focused works dedicated to individual laws or clusters:
    - a Pulsefold Codex,
    - a Resonant Gravity Codex,
    - a Resonant Thermodynamics Codex,
    - a Quamitry of Life volume, etc.

This Field Guide is meant to:

- Put your name on the core framework.
- Give you a starting point to point people to when they ask "So what is Quamitry?"
- Serve as a **bridge** between the poetic Horizon, the GCI Codex and the coming Master Codex.

CARD 45: USING THIS FIELD GUIDE

Mantra
Start where you are, use what you have, read what calls to you.

A practical way to use this book:

- If you are **a physicist or engineer** – start with Parts II, IV and VI (Foundational Laws, Fields and Forces, Instruments), then circle back to Part I for the ontology.
- If you are **a biologist or clinician** – start with Parts III and V (Time, Pulse, Life and Echo), then connect back to the core Laws.
- If you are **a curious general reader** – read Parts I and VII first to get the big picture, then dip into whatever card title pulls at you.
- If you are **a future Quamitry practitioner** – treat this as a command-line reference: a set of man pages for the universe. Keep it near the GCI Codex and whatever instruments you are building.

If Quamitry is right, then the universe has always been doing this. The Lattice has always been folding, remembering, forgetting, pulsing and echoing. This Field Guide is just one small attempt to describe that behavior cleanly enough that others can join in.

That is why it matters.

Appendix A – Canon Law & Principle Index

Prime Law – Law of Geometric Instruction –
Compression as instruction; geometry as code.
Law of the Lattice Principle – The universe is structure
before it is stuff.
Law of Preform Density – "Void" is a dense sea of GOE
and SubQUAMIs.
Law of SubQUAMI Resonance Lock – When resonance
agrees, the lattice remembers.
Law of Resonant Motion – Motion is locks trading places
along filaments.
Law of Polarized Resonance – Charge = direction of
filament tension (inward vs outward).
Law of Resonant Delay – Time = accumulated delay
between locks along strained paths.
Law of GOE Transformation – GOE moves between free
field and compressed folds; energy, mass, heat, work are
positions in that cycle.
Law of Dimensional Gradients – Effective dimension =
how many cheap directions remain for resonance.
Law of Resonant Scale Coupling – Same laws apply at all
scales; folds at one scale become nodes at the next.
Law of Resonant Gravity – Gravity = inward fold of
compression memory.
Law of Electromagnetic Resonant Geometry – Fields are
organized tension in the lattice.
Law of Electromagnetic Resonant Thermodynamics –
Heat, light, and magnetism are resonance conversations
between folds.
Law of Resonant Boundary – Edges are negotiation layers
where two fold regimes share the same lattice.
Law of Anchor Saturation – Overloaded polarized filaments
snap in a re-lock cascade: lightning, arcs, flares.
Law of Resonant Information – Information = structured,
re-instantiable differences in GOE/folds.
Law of Resonant Catastrophe – Collapse is over-success:

rare perfect alignment that forces geometry to reformat. *(plus Principles: Fit Horizon, Echoform, Resonant Encounter, Reflective/Obedient Absence, Rhythm of Time; Pillars; Pulsefold Hypothesis; etc.)*

Appendix B ~ Entanglement as Shared Geometry

Standard quantum mechanics describes entanglement as a state that cannot be factored into independent parts. Quamitry rephrases that in geometric terms:

Two distant folds are entangled when they share a coherent corridor in the lattice that constrains their lock patterns as a single object.

Key points:

- The **SubQUAMI / mirrorfield structure** linking the two sites is established during their joint preparation.
- That structure persists as a **shared boundary condition** even when the folds are separated in space.
- When one site is "measured," the lattice is not updating two separate systems. It is resolving **one corridor** into a stable lock pattern that must be consistent at both ends.

This picture preserves:

- All the observed correlations in Bell-type experiments.
- The absence of superluminal signals between local measurements.

It changes **where** the weirdness lives:

- Entanglement is no longer "information traveling faster than light."
- It is the fact that the **geometry itself** was already a single object spread across the region.

In RTI language, a mature resonance-tracing instrument could, in principle:

- Map not only local delays, but **correlated delay changes** across separated sites.
- Identify where anchor networks and filaments behave as if they belong to a single shared structure.
- Distinguish between mere statistical correlation and true **shared boundary coherence**.

A full Quamitry treatment of entanglement~including double-slit and delayed-choice scenarios~belongs in its own Codex. Here the goal is modest:

- To state that entanglement fits naturally into a picture where the lattice holds shared constraints.
- To suggest that resonance tracing is the right kind of tool to make those constraints visible.

www.ingramcontent.com/pod-product-compliance
Lightning Source LLC
Chambersburg PA
CBHW040857210326
41597CB00029B/4879